The Institute of British Geographers
Special Publications Series

28 The Challenge for Geography

The Institute of British Geographers
Special Publications Series

EDITOR: N. J. Thrift
University of Bristol

18 River Channels: Environment and Process
Edited by Keith Richards

19 Technical Change and Industrial Policy
Edited by Keith Chapman and Graham Humphrys

20 Sea-level Changes
Edited by Michael J. Tooley and Ian Shennan

21 The Changing Face of Cities: A Study of Development Cycles and Urban Form
J. W. R. Whitehand

22 Population and Disaster
Edited by John I. Clarke, Peter Curson, S. L. Kayastha and Prithvish Nag

23 Rural Change in Tropical Africa
David Siddle and Kenneth Swindell

24 Teaching Geography in Higher Education
John R. Gold, Alan Jenkins, Roger Lee, Janice Monk, Judith Riley, Ifan Shepherd and David Unwin

25 Wetlands: A Threatened Landscape
Edited by Michael Williams

26 The Making of the Urban Landscape
J. W. R. Whitehand

27 Impacts of Sea-level Rise on European Coastal Lowlands
Edited by M. J. Tooley and S. Jelgersma

28 The Challenge for Geography
Edited by R. J. Johnston

In preparation

Salt Marshes and Coastal Wetlands
Edited by D. R. Stoddart

Demographic Patterns in the Past
Edited by R. Smith

The Geography of the Retailing Industry
John A. Dawson and Leigh Sparks

Asian-Pacific Urbanization
Dean Forbes

For a complete list see pp. 248–249

The Challenge for Geography

A Changing World: A Changing Discipline

Edited by R. J. Johnston

First published 1993

Blackwell Publishers
108 Cowley Road
Oxford OX4 1JF
UK

238 Main Street, Suite 501
Cambridge, Massachusetts 02142
USA

British Library Cataloguing in Publication Data

A CIP catalogue record for this book is available from the British Library.

Library of Congress Cataloging-in-Publication Data

The Challenge for geography: a changing world, a changing discipline / edited by R. J. Johnston.
p. cm. — (The Institute of British Geographers Special publications series; 28)
Includes bibliographical references (p. x–xi) and index.
ISBN 0–631–18713–8. — ISBN 0–631–18714–6 (pbk.)
1. Geography. I. Johnston, R. J. (Ronald John) II. Series: Special publication series (Institute of British Geographers); 28.
G116.C43 1993
910—dc20 92–25239
CIP

Typeset in 11 on 13 pt Plantin
by Graphicraft Typesetters Ltd., Hong Kong
Printed in Great Britain by T.J. Press (Padstow) Ltd. Padstow, Cornwall.

This book is printed on acid-free paper

Contents

A Changing World: A Changing Discipline? An Introduction

R. J. Johnston

Geography is pre-eminently an empirical discipline, concerned with understanding the world and transmitting that understanding to a wide audience. It is also a practical discipline, for its transmitted understanding is of value to those who would change the world – at all scales. Thus change in the external world is one of the major stimuli to change in the discipline – in the theoretical apparatus on which it draws, in the research methodologies its practitioners employ, in the content of its educational curricula, and in its contributions to influencing change.

Change is continuous, though its rate and direction vary over time and space. Geographers have long been charting, understanding and creating changes; their educational mission to enhance individual and collective awareness of the world means that many geographers believe that they must continually respond to external forces if their discipline is to remain relevant and vital. To a large extent, therefore, the title of this book is no more than a statement of the obvious. The world changes: geography changes with it (as numerous historians of the subject have argued: Freeman, 1961: Taylor, 1985; Johnston, 1991).

So why have a book on that symbiotic relationship? Over the last two decades a number of volumes have appeared with titles that suggest an evaluation of geography's present and future content: they include *Directions in Geography* (Chorley, 1973), *Geography Matters!* (Massey and Allen, 1984), *The Future of Geography* (Johnston, 1985), *Horizons in Physical Geography* (Clark, Gregory and Gurnell, 1987), *Horizons in Human Geography* (Gregory and Walford, 1989), *New Models in Geography* (Peet and Thrift, 1989), *Remodelling Geography* (Macmillan, 1989), *The Power of Geography* (Wolch and Dear, 1989) and *Remaking Human Geography* (Kobayashi and Mackenzie, 1989). Surely we have had enough, and geographers would be better expending their energies 'doing geography' rather than 'writing about what geography is and what it ought to be'?

Those books have not addressed the issues tackled here, however. If the world that geographers study is changing fast – perhaps faster than ever before – then how should they respond? Can we just assume that they will, because they always have, or ought we to focus some of our attention on how they might, perhaps should? The context for that last question is crucial: formerly – when change was in general slower – the need to respond, positively and with some unity, was not as pressing as it is now. But some of those with political and economic power in the currently changing world question the need for many academic disciplines and the qualifications gained from their study, and so those disciplines' continued strength is increasingly predicated on their ability to meet external demands. They must be 'sold', though not necessarily in a crudely materialistic way: students must want to study them; funding agencies must want their research skills; and so forth. Geographers must demonstrate that their understanding of the world is knowledge that others need; people must be convinced that they want geography – that indeed they cannot do without it.

This book is built around the two main themes of its subtitle. Its first concern is to appreciate how and why the world is changing. From that appreciation stems the second concern – to evaluate whether and, if so, how the academic discipline of geography needs to be altered to accommodate that changing world. The tasks are massive, of course, and the following chapters provide only outline sketches accompanied by in-depth explorations of a few themes. The two introductory chapters (1 and 8) introduce the changes that pose the challenge; the remaining nine chapters are detailed examinations of specific salient points.

The origins of this volume are a session that I organized at the 1991 Institute of British Geographers' annual conference, under the title which is this book's subtitle. I am grateful to the conference organizers for including the session in the programme, to the ten individuals who accepted my invitation to contribute to the session and to stimulate discussion of the theme, and to the members of the audience who participated in the discussions then and afterwards. Nine of the ten contributors agreed to develop their arguments into written essays for this book, and I am extremely grateful to them for the quality of those essays and their willingness to work to the timetable set. I am particularly indebted to Peter Jackson, Ian Simmons and Susan Smith for their thorough critiques of my chapters – and totally absolve them from any responsibility for what remains. Finally, my thanks to Nigel Thrift who, as editor, has guided this volume through the procedures which govern the operation of the Institute's Special Publications series.

REFERENCES

Chorley, R. J. (ed.) 1973: *Directions in Geography*. London: Methuen.

Clark, M. J., Gregory, K. J. and Gurnell, A. M. (eds) 1987: *Horizons in Physical Geography*. London: Macmillan.

Freeman, T. W. 1961: *A Hundred Years of Geography*. London: Gerald Duckworth.

Gregory, D. and Walford, R. (eds) 1989: *Horizons in Human Geography*. London: Macmillan.

Johnston, R. J. (ed.) 1985: *The Future of Geography*. London: Methuen.

Johnston, R. J. 1991: *Geography and Geographers: Anglo-American Human Geography since 1945* (fourth edition). London: Edward Arnold.

Kobayashi, A. and Mackenzie, S. (eds) 1989: *Remaking Human Geography*. Boston, MA: Unwin Hyman.

Macmillan, B. (ed.) 1989: *Remodelling Geography*. Oxford: Basil Blackwell.

Massey, D. and Allen, J. (eds), 1984: *Geography Matters!* Cambridge: Cambridge University Press.

Peet, R. and Thrift, N. J. (eds) 1989: *New Models in Geography*. Boston, MA: Unwin Hyman.

Taylor, P. J. 1985: The value of a geographical perspective. In R. J. Johnston (ed.), *The Future of Geography*. London: Methuen, 92–110.

Wolch, J. R. and Dear, M. J. (eds) 1989: *The Power of Geography*. Boston, MA: Unwin Hyman.

PART I

A Changing World

1

A Changing World: Introducing the Challenge

R. J. Johnston

The axiom of the case for geographical study is the vast and complex mosaic comprising the earth's surface. Places differ, in a myriad ways. For several of the early decades of this century geographers ordered that diversity through the concept of the region (Johnston, 1984), using what became to some an increasingly tired and stereotyped methodology that added volumetrically to information but much less so to understanding (Johnston, 1991b). There are occasional calls to revive that approach (Hart, 1982); more common are arguments either for a 'new regional geography' (Gilbert, 1988; Pudup, 1988) or for an approach to geography which recognizes the particularities, if not singularities, of places (Entrikin, 1991; Johnston, 1991a). The framework for studying a changing world employed here is set within the last of those genres.

The outer structure of the framework comprises three components which encompass geographers' main topical interests: the physical environment; the created environment; and society. Each has many subcomponents, all of which interact both within components and across component boundaries. All vary spatially in their characteristics, and those variations produce the rich mosaic and its myriad different places whose understanding occupies the heart of geographical scholarship. In some cases, one component (or even one subcomponent) dominates in providing the defining parameters of a place, but the components' interaction is crucial in the great majority.

THE PHYSICAL ENVIRONMENT

Traditional regional studies had the physical environment as the base on which all other aspects of geography are built. The case that nature is the foundation of human survival cannot be countered, for in most places the local physical environment underpinned its inhabitants' quality of life. Unfortunately, the intellectual base for much of the work was deeply rooted in environmental determinism, a belief that nature was not just a major constraint to human activity but very largely its determinant. This naive conception of the interrelationships between humans and nature led to a very sterile, descriptive regional geography which the majority of academic geographers had disowned by the 1960s. They preferred to focus on topical specialisms, which led to a fragmentation of the discipline, not only between physical and human geography but also within each.

Much effort has been expended by physical geographers on research into how 'nature works' in the production of landforms, climates, and faunal and floral assemblages. This has largely involved separate activity by groups looking at the atmosphere, the land surface, plant and animal life, the soil, and the oceans. Stoddart (1987:329–30) has criticized this division in writing about all, and not just physical, geography:

> We call ourselves not just physical or human geographers, but biogeographers, historical geographers, economic geographers, geomorphologists. We each develop our own expertise, our own techniques, our own theoretical constructs. Necessarily so, if we are to make our mark in scholarship. It is not the fact of this specialization within the field that I object to . . . but its consequences. And the chief of these is that for too many of us the central idea of geography . . . has disappeared.

Thus although there have been major advances in understanding the parts (Gregory, 1985), there has been less achievement in assembling the whole. Some of the parts studied are very small; there are substantial subdivisions within geomorphology, for example, generating a myopia linked to a relative absence of synthesis and vision. A few have countered by inventing 'macrogemorphology', and Summerfield (1989:431–2) contends that 'In the context of increasing integration in science, the world of earth surface process geomorphology linked to hydrology but simultaneously semi-detached from the main body of earth science research seems out-dated and in need of revision': he has sought to link physical geographers with geologists and geophysicists, in order to promote the former's appreciation of the macroscale tectonic processes involved in landform development.

The study of wholes – sums of interacting parts – has been promoted through the concept of the system. Its basic conception and representation for didactic purposes are readily undertaken and appreciated – through 'box and arrow' diagrams, for example (Huggett, 1980) – but formal modelling of the myriad interactions within most systems is a mathematical task that few geographers have tackled (see Goudie in chapter 6 below). For those who have, the difficulties of appreciating how the environment works have become increasingly apparent. Advances in mathematical modelling have shown that what appear to be relatively simple relationships are actually far from that: population change within a constrained environment (a maximum food supply, for example) is not only nonlinear but is also chaotic (where chaos is defined as a sequence of values for population size which contains neither pattern nor order: see Johnston, 1989a). If the apparently simple relationship between the population of an area at one time and the population at the next, up to a prescribed maximum, is chaotic, then the multivariate interactions among many separate factors – as in the general circulation of the atmosphere and its relationships with local climate and weather patterns – provide massive challenges to understanding and its transmission, let alone to attempts at predicting the future.

Those challenges must be met, however, because of the rapidly expanding need to appreciate how nature works. But the presence of chaotic relationships and 'bifurcations' within the environment calls for sophisticated scientific skills, allied to a fertile imagination. (Bifurcations are discontinuous changes brought about because a crucial threshold is crossed – as in the transition between meandering and braiding states of rivers: see Curry's 1962:24 contention that climatic change, of the magnitude of the onset of an ice age, could have resulted 'purely from random events'.) The importance of developing those skills and of fostering those imaginations is growing because change in the physical environment is accelerating, largely, it seems, through the increased impact of human activity. As Simmons and Goudie make clear (chapters 5 and 6), although we are in no doubt that human activity is altering the physical environment at a growing rate, as yet the full extent of those alterations and their potential future impact (which may already be set in train and irreversible) cannot be judged. Some writers are thus harbingers of doom, whereas others are relatively optimistic about the future of life on the planet. The major difficulty posed by these alterations is that we may not be able to determine who are right until too late; can we afford to wait, trusting in the optimistic view? Given the predictions and forecasts that have been produced in recent years – from sober analysts, not from sensationalists – then greater efforts are needed to understand how the physical environment works (in parts and, especially, as a whole) and how those workings are influenced by what people do, so that we can be prepared for the worst-case scenarios of our potential environmental futures.

This potential focus is linked to the study of self-regulating systems referred to by Simmons in chapter 5. Such systems are characteristic of a cosmos which is far from thermodynamic equilibrium; they are open to evolutionary change at all temporal scales. A key element within them is their feedback loops. Human presence reinforces some of those loops, while removing others: both the physical-ecological and the intellectual-symbolic characteristics of *Homo sapiens* are brought into play. The variety of temporal scales involved in the accounting reminds us that *Homo sapiens* is a product of organic and cosmic evolution and is therefore both created by the past and a creator of the future (as is made clear by Marx's classic phrase that people make history, but not in conditions of their own choosing, and, more recently, by Giddens's concept of structuration: Giddens, 1984). Further, variation in temporal scales is associated with variation in spatial scales, so that the processes of creating futures in the context of the past have a range of local influences impacting on local outcomes in both time and space – though in the physical environment local impacts may spread very widely, perhaps more so than in the human.

Human activity is having a greater influence on the physical environment now than in the past, as Simmons, Goudie and Parry all make clear (chapters 5, 6 and 7) – because there are more people in the world, who want to live at ever-higher material standards, and thus are making increased demands on the earth as their resource base. Their demands and how they are expressed – in ever-wider urban sprawl and greater air and water pollution, for example – may be triggering new bifurcations and stimulating chaotic trends which fluctuate ever more widely and wildly. Investment in research into the changing physical world, the transmission of its results, and translation of their implications into policy recommendations are crucial to the changing world of the next few decades.

THE CREATED ENVIRONMENT

Investigating the created environment in this context encompasses all aspects of human modification of the environment in town and countryside. (The term 'created environment' is slightly clumsy, but it more exactly conveys the role of human creation than does either the concept of 'landscape', which carries other connotations, or the more narrowly focused term 'built environment'.) Arguments continue to rage regarding the destruction of the 'natural environment': is there any of it left, or has everything been significantly altered (indirectly if not directly) by human activity? A more important question, however, is whether current changes are more rapid than, and differ in their intensity and effects from, those which have gone before. Goudie (1986:1) placed 'whether, and to what degree, humans have during their long tenure of the earth changed it from its hypothetical pristine condition?' as one of the three basic questions concerning the people–environment relationship: his book assembles a great deal of evidence that can be used in framing an answer.

Histories of the interrelationships between humans and their environment (such as Simmons, 1989) indicate that what is often identified as 'progress' – the movement towards societies built on mass consumption of material goods – has been accompanied by greater investment in the created environment. (Harvey, 1978, argues that investment in the created environment for its own sake, rather than as part of the productive process, has become a central component of capitalism.) What are termed relatively 'simple' societies may not have always lived in a short-term symbiotic equilibrium with their physical surroundings, however. Blaikie, Brookfield and Clarke (1987:143) argue for New Guinea/Irian that:

> During this [last] 3000 years, and especially during the past 2000, we find some remarkable examples of the interplay between a land management that degrades, particularly by causing severe soil erosion, and adaptive land management, actions that adjust land use toward viable, sustained-yield agro-ecosystems. The evolutionary term 'adaptive' is used to call attention to the uncertain, halting and sometimes unexpectedly advantageous process that enables human and other organisms to move toward their future.

In the so-called 'developed industrial' societies, however, there is little doubt that 'sensitive ecosystems and fragile soils have been degraded as an interactive consequence of the witting and unwitting short-term social priorities of individuals, corporate and state institutions, and the general society' (as the situation in Australia has been described: Messer, 1987: 238): the environment has fared no better in 'socialist' countries, where the recent political changes have brought firmly to outsiders' attention the scale of the ecological disasters created during the attempts to increase the local supply of material goods both rapidly and cheaply.

Farmed land is the most extensive portion of the created environment. As pre-capitalist modes of production have been infiltrated and eventually replaced, so farmers have been impelled to increase the intensity of their demands on the land – and often not to their own medium- and long-term advantage (Watts, 1983). Goudie (1986), for example, has illustrated how interactions of population growth and the sedentarization of nomads, leading to overgrazing and increased demands for fuel and building materials, thereby stimulating tree and shrub destruction, soil erosion, and the modification of microclimates, have generated desertification in northern Kenya.

The most intensively used portion of the earth's surface is the townscape. Rapid urbanization has been a particular feature of the last two centuries only, but big cities are not recent phemonena. As Taylor (1989a) has argued, population estimates for the past seven centuries indicate how many very large cities there were prior to the period that we call the 'industrial revolution' – most of them outside the capitalist economy focused on northwestern Europe.

Cities are internally organized in a variety of complex ways: there are many differences in the details of that ordering, but also many cross-cultural similarities. The main land uses tend increasingly to be segregated, with clearly defined areas of commercial, industrial and residential land; that segregation is virtually imposed in most places today, by planning regulations designed to minimize the potential nuisances from land-use intermixture. Each of those component areas may be internally differentiated too, especially the most extensive – the residential – with its separate areas housing different socio-economic, migrant, life-style and other cultural groups. In all but a few cases residence in particular parts of a city is not prescribed to certain groups, but economic, social and cultural sorting processes interact to encourage inter-group distancing. (Apartheid in South Africa provided the best counter-example to this generalization, for there strict legal controls rigidly separated the four largest ethnic groups.) The result is a 'ghettoized' society with different groups living apart in separately defined territories and developing separate visions of society accordingly

(Johnston, 1989b, 1991a). Recent changes in society have not reduced these trends: indeed, Susan Smith (chapter 3) suggests that they may have been accentuated.

Study of land use and its spatial organization is one of the long-term strengths of human geography. As populations continue to grow rapidly, so land-use practices are altered to meet the demands for food and raw materials and to provide housing and jobs, places for interaction and areas in which to enjoy leisure-time. Many of those accommodations are relatively short-lived, as economic change makes some uses, and even some places, redundant and new layers of investment are put down. Uneven development is a characteristic of the capitalist mode of production at all spatial scales: both theoretical (Harvey, 1982; Smith, 1984) and empirical (Massey, 1984) investigations have indicated its necessity, and also the need for its spatial parameters to change. Some analysts suggest that those parameters are changing more rapidly today: what might have had a life of 50 years as a thriving industrial region a century ago may now only achieve a decade of prosperity. The intensity and pace of change are thus important foci for geographical work, as the 'created' environment is continually 'recreated'.

This rapid pace of change is both quantitative and qualitative according to some: a new era, termed either 'flexible accumulation' (Harvey and Scott, 1989) or 'disorganized capitalism' (Lash and Urry, 1987), has been inaugurated, it is claimed. The capitalist world-economy is now so integrated that events in one place rapidly resonate elsewhere; goods, people and, especially, capital are being moved rapidly from one place to another to capitalize on immediate potentials but with little intention of making long-term commitments. Harvey (1985) argued that this hypermobility requires an enhanced response by the state to promote and defend the interests of local people, and Dicken (in chapter 2) identifies trade and investment as two of the three major forces of change in the contemporary global economy.

SOCIETY

Geographers have turned their full focus on to this third component only recently: whereas most traditional regional studies concentrated on population numbers, economic activities and aspects of the created environment, few explicitly referred to the myriad ways in which a place's population both is internally differentiated and also differs from other places on the same criteria. Some recent works have corrected that imbalance, but studies of the structure of society and its interactions with the physical and created environments remain the least well-developed aspects of the discipline: Jackson (1991), for example, has argued that geographical studies of local societies continue to pay much more attention to economic characteristics than to the many other aspects of local culture, and he noted critically that for such studies places are often spatially defined as coincident with labour markets.

The relative poverty of such investigations results from geographers' failure to develop a widely accepted framework for studying spatial variations within society. The subdiscipline of cultural geography, well-established in the United States, is somewhat myopic in its focus on artefacts in the created environment rather than on the people themselves; recent developments (for example, Jackson, 1989), though encouraging, have yet to provide a blueprint (Johnston, 1991b) and we continue to explore without a good map (though see Jackson's comments in chapter 10 regarding the use of maps!). The following paragraphs sketch such a map, identifying three main components to the study of society – the sphere of production; the sphere of consumption; and the sphere of politics.

The Sphere of Production: Work and the Workplace

A recently proposed framework for geographical studies of society, and thus of spatial variations in cultural forms, draws on Weberian sociology (Urry, 1981). Three main elements have been identified (Johnston, 1991a). The first is the nature of *social relations at the workplace*, or the ways in which the world of work is organized (which includes consideration of those 'allowed' to work and those whose 'work' – as in the domestic sphere – is defined as outside 'the world of work': most studies identify the workplace with the sphere of production for goods and services only, however). Within the capitalist mode of production, for example, the fundamental relationship is that between employer and employee.

The nature of the employer–employee relationship varies substantially over time and space. The relationship that Marx portrayed in mid-nineteenth-century Britain was a direct one between capitalist and proletarian – or 'boss' and 'worker' – for many people. This was largely replaced, as the scale of organization expanded, by a relationship between corporation and proletarians – mediated on the one side by professional managers and administrators, who were responsible for investing and managing people's capital profitably, and on the other, in many cases, by trade unions and similar bodies which represented the collective interests of employees. The fundamental relationship was the same – those responsible for managing capital aimed to buy labour as cheaply as possible, within constraints, and tried to get labour to work as efficiently as possible; the difference was that the managers were now responsible for other people's money rather than their own. (Paradoxically, the money being managed might be that invested for the workers being managed – as with their pension funds!)

There has been a major shift in the nature of the employer–employee relationship in recent decades. There has been a switch in the 'regime of accumulation' – the way in which profit-making is organized – away from 'Fordist' organized capitalism, with its concentration on the mass production of goods for mass consumption, to 'flexible accumulation', with the emphasis on deconcentrated production and many niche markets. The switch is taking place unevenly over time and space, however. Thus, for example, instead of firms employing people directly in their search for efficiency in the use of labour, they have increased their subcontracting-out of activities to self-employed workers lacking permanency of employment: same goal – different method, different workforce. Home-working has increased as a consequence, and many people formerly outside the formally defined workforce in some countries (the great majority of them women) have been incorporated within the periphery of the new division of labour.

This new arrangement – characterized by Allen (1988) as 'fragmented firms, disorganized labour' – is one element in the major changes that have taken place recently in the organization of labour, which are a focus of Dicken's chapter (chapter 2). Many firms have fragmented their operations, both vertically and horizontally (or organizationally and geographically), in order to manage risk-taking and promote profit-making. Some have retained ownership of the whole but have decentralized management (with budgetary responsibility) to separate parts – many of them widely dispersed spatially; some have devolved responsibility to new small firms that are tied to them by franchising and other arrangements; and a third group have disinvested from some of their activities, preferring to rely on new small firms which are subservient to them through new forms of contractual arrangement such as 'just-in-time'. (Not all of the arrangements are new: there has been a revival of sharecropping in some agricultural regions, for example.)

This organizational and geographical fragmentation of the sphere of production, described by Thrift (1989) as growing international disorder, and the relative growth within it of service rather than manufacturing activities, has stimulated much greater volumes of international trade in capital, ideas, goods, services and people. That trade poses new challenges for states, which seek to promote the profit-making activities of businesses domiciled within their territories. As Dicken argues in chapter 2, although the relative power of state and corporation has shifted in recent years in the latter's favour, the state is still far from impotent as an influence on the geography of economic activity.

There is no single process, no set of stages, whereby one method of organizing work within capitalism succeeds another. Change occurs as people respond to situations, according to their perceptions of the constraints and opportunities presented to them. This can produce variations between places at a whole variety of scales. Within the British coal industry, for example, different methods of organizing the labour force were used in different fields at different times (Griffiths and Johnston, 1991). And, as Whatmore et al. (1987a, 1987b) have shown, capitalist social relations have infiltrated British agriculture only recently, and then at different rates and in different ways in different places. Capitalism is driven by an imperative – the accumulation of wealth through the production of goods and services for sale at a profit – but how that imperative is operated varies substantially. Different ways are developed in separate places and become part of the culture there: they are then transmitted inter-generationally and provide the contexts within which future rounds of uneven development take place, though they are always open to change as those socialized within them perceive the need for alterations (in some cases after contact with outsiders brought up accustomed to other ways of doing things).

Civil Society and the Sphere of Consumption

Outside the world of work lies the *sphere of consumption* (sometimes termed civil society) within which the labour force is reproduced both intra- and inter-generationally. The household is this sphere's basic unit in most societies, built around the nuclear family which lives in a separate dwelling, but there are many variations to this theme – in the structure of those units and in the allocation of power within them, for example, as illustrated by Todd's (1985, 1987) seminal explorations.

For some students of contemporary societies, divisions within the sphere of consumption are at least as important as those related to work and the sphere of production (as illustrated by Dunleavy's 1980 research into electoral patterns, although he later introduced a further cleavage based on public versus private ownership of major consumer goods, such as housing: Dunleavy, 1986). Saunders has written extensively on this issue (for example, Saunders, 1984, 1986), distinguishing between people whose major items of consumption (such as housing, education and transport) are obtained in the market and those who obtain them through state provision. The latter are increasingly disadvantaged, both materially and in terms of life-satisfaction, as a smaller proportion of the population remains dependent on the state (as with the growth of the owner-occupied housing market, for example) and experiences lower quality provision as a consequence. He portrays the benefits of ownership as follows (Saunders, 1985:167):

> in Britain (though not necessarily elsewhere), ownership of a home is important for many people, not only economically (in terms of the capital gains which may accrue, the potential for transfer of wealth between generations which it creates, the opportunity to use personal labour to enhance the value of the property through DIY, and so on), but also culturally and socially. Ownership of a home brings with it some limited opportunity for autonomous control – over where you choose to live, how you choose to live, and so on. The popular and widespread desire to achieve owner-occupation can in my view mainly be explained in terms of a search for a realm of individual autonomy and freedom outside of the sphere of production.

Against that, he sets the perceived lack of autonomy of those dependent on state welfare services: 'there can be no doubt that people's experience as clients of welfare agencies has been one of powerlessness in the face of administrative indifference, professional self-interest and bureaucratic insensitivity' (p. 167). Saunders thus supports policies which either increase choice within state-provided services or privatize them. (See also the discussion of Bennett's arguments regarding post-welfare societies in chapter 8.)

Within the sphere of consumption, Saunders's work focuses on major items of social reproduction, such as housing and education, and the contrast between state and market provision. Warde (1990) argues that the market and the state are not the only sources of those services which enable reproduction of labour power, however: there is also household labour, contracted-for mutual support, reliance on neighbour and kinship networks, and some employer provision. How these non-market sources operate reflects the internal structuring of societies. Individuals and households belong to social groups which form part of the structuring of their lives and have meaning for them. Not all groups are salient in every place, depending on local societal development. Gender, ethnic identification, language and religion are among the most important. Most societies have clear gender divisions: very many have ethnic divisions too, but linguistic and religious differentiation are not characteristic of all. Where some of these are extant and important, they influence various aspects of society's operations, such as access to certain jobs and residential areas: full understanding of a society and its geography is impossible without their appreciation.

As with other aspects of society, these divisions are not fixed. Some change through being successfully contested by those suffering relative deprivation – as with the many struggles for civil rights by women and by ethnic, linguistic and religious groups (usually minorities). These struggles often overlap into the other spheres, as with those over the right to work. As society changes so does its internal structuring, as both cause and effect, and covering a much wider range of activities than the items of consumption that Saunders considers.

Despite many expressions of optimism by social commentators, politicians and academics, especially in the 1960s, change in contemporary societies, including the most affluent, has not produced equality – as Susan Smith argues in chapter 3. Indeed, she contends that alongside the substantial change of recent decades there has also been considerable continuity, of inequalities and injustices within society. Further, as she illustrates, their creation and maintenance is enhanced by territorial strategies and ideologically promoted by arguments based on natural science metaphors. Cases for segregating people on gender, race, age and health grounds are all examined to show how the appeal to 'scientific' arguments is used to create and sustain geographies of discrimination and disadvantage, which contribute to the continued reproduction of an unequal world.

Susan Smith's examples focus on processes and patterns of social and spatial discrimination within cities of western societies, but similar arguments can be used at other scales. Immigration policies in many countries are designed to exclude certain racial groups, and also age groups in a number of cases, for example, using natural science metaphors (as with the health status of potential immigrants) to justify such discrimination. In a world of economic and political unequals, social inequality is advanced through a range of strategies, many of which involve the use of space, and especially bounded space, to promote the interests of some over those of others (as argued in more detail in Johnston, 1991a).

There are strong interactions between the spheres of production and consumption, because the power of individuals in the latter is very much a function of their relative success in the former. In market societies, the formal links between the two take place in the sphere of exchange, where individuals compete for privately produced goods and for jobs – that is, in both commodity markets and labour markets. Operation of those markets, including their spatial variability, is often a contributor to social and economic inequality within society, which stimulates calls for intervention by those operating in the third sphere – that of politics.

Struggle and Politics

Struggles are endemic to the spheres of consumption and production in a capitalist economy, as individuals and groups seek to improve their absolute and relative well-being. These struggles may take place in the *sphere of politics*, which has two separate, though linked, parts. A major locus of struggle in the sphere of production is at the workplace, involving workers organized into trade unions in conflict with their employers over conditions of employment, especially remuneration. Regulation of that conflict invariably involves the state apparatus, operating to ensure relative peace and so promote profit-making and relative prosperity for all. In many countries, conflict over that regulation occurs within the state too, with political parties organized to represent the different sides. Elsewhere, parties have not been successfully organized to mobilise the conflicting groups. The United States is a good example of this: the system of labour relations created there since the 1930s restricts conflicts between unions and employers to individual workplaces only (Clark, 1988) and mass political protest over working conditions is rare (Piven and Cloward, 1977).

Alongside conflicts between employers and employees are occasional differences within the former group – over access to resources, for example – which may also be mediated within the state apparatus: many countries have experienced conflicts between advocates and opponents of free trade, for example (Taylor, 1984). One of the state's main roles in capitalist societies is to promote wealth creation through profit-making, which involves choosing which industries to promote, which to protect, and which futures to leave to the unfettered operation of market forces. From the 1930s to the 1980s, the state in many countries protected a wide range of industries in the belief that this was the best way to ensure national prosperity. In the last decade this has increasingly been condemned as inefficient and protection has been removed, thereby opening up entire national economies to global forces (as in New Zealand: Johnston, 1992a).

The state is also the focus of conflict resolution in the spheres of consumption and exchange, over the provision of welfare services, for example, and the roles which individuals can play within society. After the major world recession of the 1920s and 1930s, a coherent welfare state structure was created in many countries out of a range of *ad hoc* measures designed to reduce the impact of poverty and deprivation (the United States was a conspicuous exception among the advanced industrial countries): this was the state response to 'poor people's struggles' (Piven and Cloward, 1971, 1977). But by the 1970s and 1980s this welfare state structure was widely seen as a constraint to capitalist progress and a barrier to recovery from the contemporary economic recession. The welfare state was presented as a hindrance to entrepreneurial activity, because its costs required high rates of taxation which discouraged initiative and risk-taking. It was also believed to create a culture of dependency which was a work disincentive, and the power that trade unions had won in bargaining for workers' rights was held to be akin to that of a monopoly, with all the inherent inefficiencies of such structures. A strong ideological case was presented for the reduction and restructuring of the welfare state, and the privatization (or recommodification) of many public services (Johnston, 1992a).

Although all states exist to undertake certain necessary tasks within capitalism (and also within socialism), there are many variations among states in the interpretation of how to perform those tasks. Thus the nature of struggle varies over time and space, at all scales. Regarding the accountability of the state, for example, the nature of the allocation of power differs from the extremes of democratic participation (exemplified by Switzerland) to those of centralized autocracy (as illustrated by several of the 'Gulf states').

The state apparatus is one of many focal points of power within capitalist societies. The four major ways of mobilizing power (economic, military, political and ideological) are concentrated there: no other focus combines all four, and states reserve a virtual monopoly over military power to themselves. Although power is in part exercised merely for the gratification of those involved, most of it is used to affect what happens in the spheres of production, consumption and exchange. The interactions among state, civil society and the world of work are among the defining parameters of a modern society.

As the other spheres change so the state must respond to and stimulate alterations at workplaces and within civil society, as well as in the arena of exchange. Given the rapidity of recent change within the others, therefore, it is not surprising that the state has altered rapidly in recent decades also. Some analysts have identified major changes occurring approximately once every half-century, linked to the redundancy that occurs in the investments that have created the current pattern of uneven development. These cycles – called Kondratieffs – describe the major periods of prosperity and recession that characterize the capitalist world-economy: each new cycle tends to be dominated by a new set of industries based on a new set of products, by new ways of organizing production, by new relationships within the sphere of consumption, by new centres of economic prosperity (the cores of the contemporary world-economy) which exploit the less-prosperous peripheries, by new patterns of global geopolitics (see Taylor, 1989b, 1992) and by new approaches to regulating struggle in civil society (Johnston, 1992a).

The state both responds to these waves in the world-economy and contributes to the creation of their specific geographies. Much of the response comes in the period around the low point of each cycle, when the state is called upon: (1) to assist those who have previously contributed to the creation of wealth but are currently suffering from market decline, by aiding the creation of new products (both goods and services) and the opening up of new markets; (2) to provide succour for those in the workforce who are unemployed because of the recession, providing welfare payments until employment becomes available again, plus help in obtaining training for the new types of job in the emerging industries of the next cycle; and (3) to ensure consensus within society by stimulating a set of 'ruling ideas' which interpret the current difficulties to the population in ways that will win their support for state policies during the period of difficulty (see Johnston, 1991a, on Britain under Thatcherism, and Johnston, 1992a, on similar processes in New Zealand, Germany and the United States). How those controlling the state apparatus choose to pursue such difficult, and to some extent mutually antagonistic, roles depends on their perceptions, which in turn are strongly coloured by the local cultural context in which they operate.

The nature of the state in the capitalist world-economy is rapidly changing in the wake of criticisms of its regulatory and welfare ideology, therefore; its changes are having substantial impacts on the spheres of production and consumption, as regulations are withdrawn (on the freedom of capital movements by the UK government in 1979, for example), protective policies are reduced (as with the contraction of the UK regional aid programme) and welfare provision is reduced (as with the sale of UK state housing and the very substantial reduction in the rate of construction of new stock). The state is intimately implicated in the restructuring of the spheres of production and consumption, as it attempts to ensure a new cycle of capitalist prosperity – in large part by placing more of society's activities in the markets of the sphere of exchange, which involves the state withdrawing somewhat from its role as a link between the spheres of production and consumption.

In the socialist world, the failure to deliver high material standards to the majority of the population, while at the same time restricting their civil rights and allowing little direct accountability of those exercising power, has led to the recent wholesale restructuring of the state apparatus. This has occurred in large part as a response to popular demands made possible because of the withdrawal of Soviet Union sustenance for the state apparatus in adjacent countries; the Soviet Union's state also found that the failure of its socialist policies constrained its ability both to satisfy its population and to repress the populations of satellite countries, to such an extent that the monolithic 'federation' of the USSR has been dismantled, and the nature of its successor state(s) is far from certain. The ex-socialist states are seeking to join the capitalist world-economy at a time when market forces are being freed elsewhere. Whether new countries can survive in such a context, or whether they will need to introduce protective policies of the sort being abandoned elsewhere, is uncertain: some commentators suggest that it will take 8–10 years for a free-market economy to be established successfully in Russia, for example, but such a long period may be unacceptable politically to those who want early economic prosperity to go alongside their hard-won political freedoms.

It is not only the nature of the state that is changing but also the pattern of states; the map of political boundaries has been redrawn in some places and attempts at redrawing it have been made in many more. As the map of uneven development changes, so does the political map, with geopolitical power following economic power. Empires have waxed and waned, spheres of influence expanded and contracted, states have been annexed, and so forth (Taylor, 1989b). The relative economic decline of first the UK and then the USSR and the USA during the late twentieth century has been countered by the economic growth of Germany and Japan. Both were defeated in the 1940s, and as a consequence were unable to station armed forces outside their borders. They put more of their effort into economic and social rather than defence policies, therefore (although Japan has one of the largest home defence forces in the world), and are now challenging for global economic dominance (in Germany's case, very largely through dominance of the European Community).

Recent decades have been characterized by two particular geopolitical trends which appear to move in contrary, even conflicting, directions. The first has been the creation of federations of states, to an extent not witnessed before (apart, perhaps, from the voluntary creation of federations in Australia, Canada, South Africa and the United States of America). Inter-state arrangements after each of the twentieth-century world wars were attempts to prevent further armed conflicts via international mediation and policing. The League of Nations created after World War I had that single purpose only, and conspicuously failed. The United Nations, created in 1945 and with a much larger membership, adopted a wider mandate and established a number of other agencies concerned with economic and social as well as political issues: alongside it have been regional organizations designed to promote both economic (as with the EC and COMECON) and military goals (NATO and the Warsaw Pact, for example). Most individual states have relinquished very little of their sovereignty as a consequence of membership of these bodies, however, unless coerced into so doing. Where the potential for loss of sovereignty has been perceived, as with the draft treaties on the Law of the Sea, some states, notably those with most to lose, have been reluctant to conform (Johnston, 1989a, 1992c).

Alongside the UN and its various agencies, the most important new development by far has been the extension (in both membership and powers) of the European Community. Initially established as a trading bloc of adjacent states, seeking to enhance markets and engage in collective protective practices, the EC has moved rapidly in the last decade (too rapidly for some member-states, notably the UK) towards a federal structure within which the individual members will yield substantial sovereignty to the supranational organization. If this move is sustained, a new state form will have been created.

The second trend is the resurgence of nationalist and self-determination calls from peoples long subject to the power of larger states. Whereas the trend towards larger groupings is redrawing the political map by devaluing, if not removing, some inter-state borders, this latter trend is seeking to redraw it by introducing new boundaries and upgrading the importance of others. The trends are not necessarily contradictory – some Scottish nationalists argue for a separate Scottish state within a federal Europe, for example – but they indicate that a complex reorganization is currently taking place.

Nationalism is not new, nor is it necessarily the case, as some argue, that nationalist movements are strongest in the peripheral regions of unevenly developed countries (such as Scotland and Wales within the UK) during periods of economic recession. There are many reasons for growth of nationalist sentiment, and in some cases – as in Belgium in recent decades – the calls for separate statehood draw on a latent base (linguistic differences in that example) which had not previously been the mobilizing focus for political movements. Two global trends – the increased scale of economic operations and the growing homogenization of many aspects of culture – may be the stimuli for reactions based on the 'small is beautiful' principle, however. Increasingly people see the control of events moving beyond their reach and they are therefore receptive to attempts at mobilizing them politically around the cultural base of a national identity focused on a relatively small and well-defined territory.

There are two main types of nationalist mobilization. The first involves no proposed changes to the form of the political map: the goal is to mobilize people behind a state ideology which associates them with a predefined territory, and which may involve them jettisoning previous national identities. The best examples are most of the recently independent ex-colonial states (especially in Africa) where the political boundaries were drawn by imperialist conquerors with little or no respect for the territorial claims and aspirations of the existing inhabitants. With independence, the potential for returning to older territorial boundaries is extremely limited, and politicians use the sense of community developed during the anti-colonial conflict as the foundation for a new nationalism. But such ideological developments are not confined to those countries: strong nationalist programmes within existing state boundaries were used by de Gaulle when he rebuilt French morale and sense of identity after 1958, and also by Thatcher in her goal of 'making Britain great again' in the 1980s.

The second type of nationalism involves attempts to change the political map by, for example: (1) one or more states breaking away from an existing one – as with the claims for Catalonia and Galicia in Spain and Corsica in France; (2) parts of several states being brought together to form a new one – as with the claims of the Basques in France and Spain and the Kurds in Iraq, Iran and Turkey; and (3) the removal of part of one state to join another – as with the attempts to reunite Northern Ireland with the Republic of Ireland. In all three, the nationalists claim identification with a particular territory, over which they want sovereignty. Their separate identity is usually, though not necessarily, associated with other cultural traits, such as language, religion and ethnic origin: Conversi (1990) suggests from a comparative analysis of the

Catalan and Basque cases that language provides a better foundation than race for peaceful mobilization of a national population.

The recent growth of this second type of nationalist movement is testimony to the strength of enduring separate identities associated with a territory. In some cases there have been decades, even centuries, of attempts to remove those identities – as with the anglicization policies pursued in Scotland, Wales and Ireland. These appear largely to have failed. Nowhere is this clearer at present than in the remnants of the Soviet Union, where many nations are grasping the opportunity offered by the economic difficulties of the federal socialist state to press for their own separate, independently governed future.

In some parts of the world, therefore, the cultural mosaic is being overridden, if not obliterated, by nationalisms associated with the political boundaries imposed by the colonial powers – not without substantial difficulties in some cases, because of the problems of integrating separate nations ('tribes' as Europeans prefer to call them), as in Nigeria and Zimbabwe. Elsewhere, as in much of Europe, the long-established cultural mosaic has not been obliterated by the political machinations of the last two centuries and the impacts of two world wars, and different cultural groups, as in Yugoslavia, are seeking the separate nation-state status that many of them had before, even if few of those alive today can recall it.

The changing political map is clearly linked to the changing economic and social circumstances. The break-up of the Soviet Union, for example, is a consequence of that country's failure to deliver substantial and widespread material benefits, plus the growing inability of the state apparatus to repress the deprived populations. In part, the failure and inability are linked to the priority placed on military expenditure by successive generations occupying the state apparatus, both to protect the Soviet Union from a perceived threat from the capitalist world and to sustain 'imperialist ambitions' associated with spreading the Communist ideology and way of life. The state was economically overstretched, with geopolitical ambitions that its economic performance could not sustain. Restructuring was thus 'forced upon' the leadership.

Much of the Soviet Union's fear of the 'west' focused on the United States which, as the most successful capitalist economy in the middle decades of the twentieth century, had invested much in a military machine intended to 'defend the free world' and its capitalist ideology against the Communist threat. But, as in the Soviet Union, the economic demands of sustaining that military presence worldwide, and the ever more sophisticated technology and weapons of destruction associated with it, became increasingly burdensome on the American economy. It too was overstretched because of its military commitments, and in the late 1980s its government was prepared to negotiate reductions in the arms race with the Soviets: without that, it was clearly going to lose in the contest for economic dominance of the capitalist world-economy with the Germans and Japanese. The 'August 1991 revolution' in the Soviet Union provided the context for a rapid quickening of the scaling down of commitment to the arms race on both sides, as the abandonment of Communism removed much of the perceived threat.

The 1990s, as Graham Smith illustrates in chapter 4, have been heralded by many world political leaders as the beginning of a 'new world order'. For the first time in some 45 years, the world will not be dominated by two major economic powers with separate political ideologies, using military strength to advance their causes internationally. The consequence of this, in many countries and not just the two former 'superpowers', could be either a redistribution of state finances towards economic and social goals (allocating what has become known as the 'peace dividend') or a further slimming down of the state apparatus as the reduced spending brings lower taxes and greater 'market freedom'. But although the potential for global conflict focused on the USA and USSR has been substantially reduced, that for local conflict has perhaps been enhanced. The 'Gulf War' of 1990–1 illustrates the ability of certain states (distinguished perhaps by size, relative affluence and leadership qualities) to amass large enough military machines with which to threaten neighbours, and the conflicts in Yugoslavia indicate the potential for intra-state warfare which external powers are apparently unable to prevent. The 'new world order' may herald a new global geopolitics of relative security, therefore, but also introduce a wider range of local and regional conflicts in which the resort to arms is common.

INTERACTIONS AND THE CREATION OF PLACES

Much is changing rapidly at present, therefore, as the world enters a new cycle of the capitalist world-economy and most of the countries that opted (as far as they could) to sit outside that world-economy seek to rejoin it. As societies change, so do their interactions with the created and physical environments – alongside the changing interactions among state apparatus, sphere of production and civil society. Those interactions create a great deal of variety across the earth's surface. Some recent trends in geographic thought focus on that variety and the contention that it, and its many component parts, cannot be appreciated through the development of 'grand theory': postmodernists celebrate difference, but others – such as Susan Smith in chapter 3 – argue that there are continuities and consistencies which provide a sense of order to the world (or at least to many people's views of the world), and which call for an ordered response from analysts.

A further trend within geography in recent years, building on and contributing to a broader strand in other disciplines and in society at large, has been the regrowth of environmental awareness. There is a greatly enhanced appreciation that the physical world is being substantially altered, and perhaps irreparably harmed if not destroyed, by human attitudes to nature which are in the main exploitative and short-term rather than conservative and long-term. (Some argue that the shift away from the welfare state exacerbates those problems as state-imposed protection of the environment is abandoned to the free market.) In the heady days of the late 1960s and 1970s, when it still seemed that the welfarist paradigm could guarantee continued prosperity, Inglehart (1977) argued that people's political imperatives would shift from issues of wealth creation and distribution to those of 'new social movements' such as environmentalism. As the weight of scientific argument regarding the harm being done to the planet's life-support systems grew, so more people joined environmental movements and many others were generally convinced by the need for 'green policies'. But with the recession that followed, people, and especially those suffering most from the recessions's impact, turned again to the economic issues: the foundation of support for the 'green cause' appears to be there, in Britain at least, but it has been demoted on the political agenda relative to other issues (Johnston, Russell and Pattie, 1991).

The interactions among the physical environment, the created environment and society (each with its own separate interacting components) create the subject matter of geography – places. Places have identities created by the people who occupy them, and others created for them by those who live elsewhere. People learn who and what they are in a place, being influenced both directly and indirectly by the others who live there. The nature of those influences partly reflects the physical and created environments which provide the enveloping context within which lives are structured. As people learn and act, so they in turn affect the nature of the physical environment and restructure the created.

Life is a continuing dialectic between peoples and places, which proceeds over a range of spatial scales from the global to the very local. The outcome at the former level is the map of uneven development, in which the role and function of a place in the global division of labour strongly influences the quality of life and indeed the life-chances of the great majority of people there. Places on the periphery of the world-economy, whose resources (physical and human) are largely exploited for the benefit of people living elsewhere, are much more likely to live in deprived circumstances, in squalid homes and neighbourhoods surrounded by devastated physical environments, than are those in the core areas, who benefit appreciably (if indirectly, in the great majority of cases) from the exploitation of the periphery. And the former people are likely to be born with much lower life expectancies, to enjoy much inferior human and civil rights, and so forth (Johnston, 1989c). To survive, many are pressed to make ever-increasing demands on the physical environment to produce materials that can be exported to the core (at ever-falling real prices): if they do not, or cannot because the environment no longer responds to their demands, then their life-chances will suffer. In most parts of the world, but especially in its economic, social and political peripheries, the dialectic between people and environment is hastening the long-term deterioration, if not the permanent destruction, of vast tracts of land.

This trend is not a new one, as Simmons points out in chapter 5, but it is accelerating, and that acceleration is promoting a new physical geography of the earth as, for example, climatic zones are shifted in response to global warming: the implications of those shifts are profound, as Parry stresses in chapter 7. Understanding why and how the trend is accelerating is crucial to the long-term future human occupancy of the earth – and certainly so if the levels of material affluence enjoyed by a small proportion of the world's population are to be both maintained and spread. But can we be sure that is the case? We have plenty of evidence of increasingly desertified landscapes, of areas blighted by industrial pollutants, and of changes to the atmosphere which will have climatic consequences of substantial magnitude. Much of the scientific literature is still equivocal, however, perhaps for no other reason than that our understanding of a very complex situation is at best incomplete and at worst rudimentary. Yet, assuming that at least some of the best evidence currently available is correct, then if we wait for much better we may well have waited too long. The onus is on scientists to convince political and other decision-makers that the likely environmental impacts of current trends are so life-threatening that steps must be taken immediately to reverse them – and those steps must involve determined action now (Johnston, 1992c).

Society–environment interactions have long been a central concern of geographical scholarship – or so we often claim. And yet, Andrew Goudie argues (chapter 6) that geographers are doing relatively little on this increasingly crucial topic. More research is called for on the causes and nature of environmental change if we are to ensure a future that will be sympathetic to human needs. Alongside it must go the development of an appreciation of the current constraints to the imposition of such control (Johnston, 1989a, 1992b). The self-regulating free market so popular with many politicians at present may hasten rather than prevent the wholesale destruction of the physical environment.

At the global scale, the maps of uneven development and of environmental deterioration are perhaps closely correlated – which is not to argue that such deterioration is absent from the core areas of the world-economy. At the local scale, the correlation is between uneven development and the built portions of the created environment. As Susan Smith argues in chapter 3, space is implicated in the creation and maintenance of inequalities within society: the rich and the poor, the oppressed and the comfortable, the sick and the well tend to be segregated into separate parts of cities. Urbanization has been a (probably necessary) consequence of the search for material advancement under both capitalism and socialism, sucking hundreds of millions of people into dense concentrations of totally 'artificial' environments. And as the search for greater material affluence continues, so more are sucked in to cities and more is invested in the created environment, not just in providing machines for urban living but also as means of realizing wealth from the markets in land, homes, transport and the commodities to put in those homes (Harvey, 1978). The urban landscape is spreading rapidly too, as more people seek to escape the highest density concentrations: thus more of the physical environment is threatened with permanent transformation to an essentially unproductive state.

One of the many ways in which people structure their lives within the created urban environment is by defining boundaries and representing them in the landscape. This operates at all scales, from the child in a school dormitory through to the largest state. In each case, boundaries are demarcated and defended (a process of territoriality which is characteristic of other species also, but which is rarely used as an analogy for the understanding of human spatial behaviours: Sack, 1983, 1986; Johnston, 1989b). Places with which they can identify help people to gain a sense of self-identity, safety and security in an uncertain world. They provide a manageable scale of meaning, which breeds localistic feelings. And perhaps, as the processes that govern our lives become more global in scale, so the need for localism to counter it will increase: places may become more important in people's daily lives, and the desire to protect spatial differences more intense.

The earth has thus been transformed into a 'world of containers', and the conflict between the residents of these various containers – on all spatial scales – is fundamental to societal operations. Through those conflicts, the physical environment is modified (as illustrated by the ecological consequences of the 1991 'Gulf War') and the created environment reorganized. Conflict between people defending territories is central to the creation and recreation of geographies (Johnston, 1991b, 1991c).

CHARTING THE CHANGES

The preceding paragraphs have developed the challenge that understanding a rapidly changing world presents to geographers, and have provided a framework for appreciating the nature of those changes. The next six chapters extend the argument, focusing on particular aspects of the challenge in some depth. In such a brief span, it is impossible for the individual authors to do more than introduce some important themes in the challenge of a changing world which faces geography and geographers.

In the first three chapters, Peter Dicken, Susan Smith and Graham Smith develop themes related to the three components of the study of society set out above. Dicken concentrates on the world of work, with particular reference to the geographies of production, trade and foreign investment: his comments on the role of the state in the regulation of these three geographies, through policies on business organization, investment and trade, illustrate the important changing interactions between the worlds of work and politics, and illustrate how the details of the new world-economy will be worked out according to the policies of people in particular places.

Susan Smith's focus is on civil society, and in particular on inequalities among groups in their access to the benefits of material affluence. As Dicken does, she too shows the important roles of the state, both ideologically, in providing support for certain arguments (the natural science metaphors underpinning views about gender, ethnicity and health), and instrumentally, through the production of spatial boundaries to the allocation of welfare benefits of various kinds. These processes are not confined to the sphere of consumption, but apply to those of production and politics too.

In the final chapter of this trilogy, Graham Smith writes about one aspect of the major contemporary changes occurring within the sphere of politics – the collapse of states whose ideologies and policies were dominated by 'actually existing socialism'. Understanding what has happened in the Soviet Union and its former satellite states in recent years is crucial not only to an appreciation of the present and the likely future there but also to an assessment of likely geopolitical futures. The collapse of 'actually existing socialism' affects the economy and politics of the entire world, with the end of the Cold War and the development of new economic, social and political interrelationships between the formerly socialist countries and the rest of the world. These new relationships are being forged while the political map of much of the Eurasian continent east of the former Iron Curtain is being contested and redrawn – processes which contain within them the potential for much violent conflict into which, as in Yugoslavia, other countries will almost certainly be drawn. We are passing through what Smith calls, after Taylor (1991), a 'geopolitical transition' which, he notes, contains a geographical paradox: it incorporates both the need for the formerly socialist economies to be integrated with those of the 'capitalist west' that they want to join, and the desire of many nationalist groups within those countries to create separate, smaller and independent political entities.

The cases made in chapters 2–4 leave us in no doubt that the human geography of the world is changing, and changing rapidly. So too is the physical geography, which is the topic of the next three. The discussion here is holistic: rather than divide the changing physical world into its separate realms, the three authors focus on different time-scales. Ian Simmons begins (chapter 5) by looking at the long view, at the interrelationships between societies and their environments over the last ten millennia. He argues that very little of what might be termed 'pristine nature' now exists, because almost all ecosystems have been substantially changed by human action, increasingly so in recent times.

Andrew Goudie follows with a detailed evaluation of land transformation over the last few decades of intensive human modifications to environmental processes (chapter 6) and an exploration of the likely changes in the next few. In the light of the evidence unearthed, he then critically examines geographers' performance in seeking to understand those trends and their impact, and finds it wanting. His call for greater geographical involvement in understanding the new global physical geography that is currently being created, and which will have major repercussions on human geography too, is followed by Martin Parry's discussion of climate change and its impact. How do we develop models that will allow us to appreciate both the physical and the human consequences of the changes that we are currently stimulating, and how do we react to their predictions? Do we seek to prevent, or at least reduce, the changes in some way, or do we rather seek to accommodate the changes by developing new geographies of production and consumption? (Or, as Zelinsky, 1970, suggested on a related topic two decades ago, do we either 'ignore the problem, hoping it will go away' or assume that it will never really be that bad?)

Those questions lead us towards the subject matter of the second part of this book – the role of geographers as they face the challenge of the changing world. Chapters 5–7 further outline the nature of that changing world, and make clear just how great the challenge will be.

REFERENCES

Allen, J. 1988: Fragmented firms, disorganized labour? In J. Allen and D. Massey (eds), *The Economy in Question*, London: Sage Publications, 184–228.

Blaikie, P. M., Brookfield, H. C. and Clarke, W. 1987: Degradation and adaptive land management in the ancient Pacific. In P. M. Blaikie and H. C. Brookfield (eds), *Land Degradation and Society*, London: Methuen, 143–5.

Clark, G. L. 1988: *Unions and Communities under Siege*. Cambridge: University of Cambridge Press.

Conversi, D. 1990: Language or race? *Ethnic and Racial Studies*, 13, 50–70.

Curry, L. 1962: Climatic change as a random series. *Annals of the Association of American Geographers*, 52, 21–31.

Dunleavy, P. J. 1980: *Urban Political Analysis: the politics of collective consumption*. London: Macmillan.

Dunleavy, P. J. 1986: The growth of sectoral cleavages and the stabilisation of state expenditures. *Environment and Planning D: Society and Space*, 4, 129–44.

Entrikin, J. N. 1991: *The Betweenness of Place*. London: Macmillan.

Giddens, A. 1984: *The Constitution of Society*. Cambridge: Polity Press.

Gilbert, A. 1988: The new regional geography in English- and French-speaking countries. *Progress in Human Geography*, 12, 208–28.

Goudie, A. S. 1986: *The Human Impact on the Natural Environment* (second edition). Oxford: Basil Blackwell.

Gregory, K. J. 1985: *The Nature of Physical Geography*. London: Edward Arnold.

Griffiths, M. J. and Johnston, R. J. 1991: What's in a place?. *Antipode*, 23, 185–213.

Hart, J. F. 1982: The highest form of the geographer's art. *Annals of the Association of American Geographers*, 72, 1–29.

Harvey, D. 1978: The urban process under capitalism. *International Journal of Urban and Regional Research*, 2, 101–32.

Harvey, D. 1982: *The Limits to Capital*. Oxford: Basil Blackwell.

Harvey, D. 1985: The geopolitics of capitalism. In D. Gregory and J. Urry (eds), *Social Relations and Spatial Structures*, London: Macmillan, 128–63.

Harvey, D. and Scott, A. 1989: The practice of human geography: theoretical and empirical specificity in the transition from Fordism to flescible accumulation. In B. Macmillan (ed.), *Remodelling Geography*, Oxford: Basil Blackwell, 217–29.

Huggett, R. 1980: *Systems Analysis in Geography*. Oxford: Clarendon Press.

Inglehart, R. 1977: *The Silent Revolution*. Princeton, NJ: Princeton University Press.

Jackson, P. 1989: *Maps of Meaning*. London: Unwin Hyman.

Jackson, P. 1991: Mapping meanings: a cultural critique of locality studies. *Environment and Planning A*, 23, 215–28.

Johnston, R. J. 1984: The region in twentieth century British geography. *History of Geography Newsletter*, 4, 26–35.

Johnston, R. J. 1989a: *Environmental Problems: nature, economy and the state*. London: Belhaven Press.

Johnston, R. J. 1989b: People and places in the behavioural environment. In F. W. Boal and D. N. Livingstone (eds), *The Behavioural Environment*, London: Routledge, 235–52.

Johnston, R. J. 1989c: The individual and the world-economy. In R. J. Johnston and P. J. Taylor (eds), *A World in Crisis?: geographical perspectives* (second edition), Oxford: Basil Blackwell, 200–28.

Johnston, R. J. 1991a: *A Question of Place*. Oxford: Basil Blackwell.

Johnston, R. J. 1991b: *Geography and Geographers* (fourth edition). London: Edward Arnold.

Johnston, R. J. 1991c: Territoriality and the state. In G. B. Benko (ed.), *Territoriality and the Social Sciences*, Ottawa: University of Ottawa Press.

Johnston, R. J. 1992a: The internal operations of the state. In P. J. Taylor (ed.), *The Political Geography of the Twentieth Century*, London: Belhaven Press, 000–00.

Johnston, R. J. 1992b: Markets, states and the environment. *Integrated Environmental Management*, 5, 22–3.

Johnston, R. J. 1992c: Laws, states and super-states. *Applied Geography*, 12, 000–00.

Johnston, R. J., Russell, A. T. and Pattie, C. J. 1991: Is Britain going green?. *Journal of Rural Studies*, 7, 285–98.

Lash, S. and Urry, J. 1987: *The End of Organized Capitalism*. Oxford: Polity Press.

Massey, D. 1984: *Spatial Divisions of Labour*. London: Macmillan.

Messer, J. 1987: The sociology and politics of land degradation in Australia. In P. M. Blaikie and H. C. Brookfield (eds), *Land Degradation and Society*, London: Methuen, 232–8.

Piven, F. F. and Cloward, R. A. 1971: *Regulating the Poor*. New York: Vintage Books.

Piven, F. F. and Cloward, R. A. 1977: *Poor People's Movements: why they succeed, how they fail*. New York: Random House.

Pudup, M.-B. 1988: Arguments within regional geography. *Progress in Human Geography*, 12, 369–90.

Sack, R. D. 1983: Human territoriality: a theory. *Annals of the Association of American Geographers*, 73, 55–74.

Sack, R. D. 1986: *Human Territoriality: its theory and history*. Cambridge: Cambridge University Press.

Saunders, P. 1984: Beyond housing classes: the sociological significance of private property rights in means of consumption. *International Journal of Urban and Regional Research*, 8, 202–25.

Saunders, P. 1985: The new right is half right. In A. Seldon (ed.), *The New Right Enlightenment*, London: Economic and Literary Books, 163–72.

Saunders, P. 1986: *Social Theory and the Urban Question* (second edition). London: Hutchinson.

Simmons, I. G. 1989: *Changing the Face of the Earth*. Oxford: Basil Blackwell.

Smith, N. 1984: *Uneven Development: nature, capital and the production of space*. Oxford: Basil Blackwell.

Stoddart, D. R. 1987: To claim the high ground: geography for the end of the century. *Transactions, Institute of British Geographers*, NS 12, 327–36.

Summerfield, M. A. 1989: Tectonic geomorphology: convergent plate boundaries, passive continental margins and supercontinent cycles. *Progress in Physical Geography*, 13, 431–41.

Taylor, P. J. 1984: Accumulation, legitimation and the electoral geographies within liberal democracy. In P. J. Taylor and J. W. House (eds), *Political Geography: recent advances and future directions*. London: Croom Helm, 117–32.

Taylor, P. J. 1989a: The error of developmentalism. In D. Gregory and R. Walford (eds), *Horizons in Human Geography*, London: Macmillan, 303–19.

Taylor, P. J. 1989b: *Political Geography: world-economy, nation-state and locality* (second edition). London: Longman.

Taylor, P. J. 1991: *Geopolitical Transition*. London: Belhaven Press.

Taylor, P. J. 1992: Tribulations of transition. *The Professional Geographer*, 44, 10–13.

Thrift, N. J. 1989. The geography of international economic disorder. In R. J. Johnston and P. J. Taylor (eds), *A World in Crisis?: geographical perspectives* (second edition). Oxford: Basil Blackwell, 16–78.

Todd, E. 1985: *The Explanation of Ideology*. Oxford: Basil Blackwell.

Todd, E. 1987: *The Cause of Progress*. Oxford: Basil Blackwell.

Urry, J. 1981: *The Anatomy of Capitalist Societies*. London: Macmillan.

Warde, A. 1990: Production, consumption and social change: reservations regarding Peter Saunders' sociology of consumption. *International Journal of Urban and Regional Research*, 14, 228–48.

Watts, M. 1983: *Silent Violence*. Berkeley, CA: University of California Press.

Whatmore, S., Munton, R., Little, J. and Marsden, T. K. 1987a: Towards a typology of farm businesses in contemporary British agriculture. *Sociologia Ruralis*, 27, 21–37.

Whatmore, S., Munton, R., Little, J. and Marsden, T. K. 1987b: Interpreting a relational typology of farm businesses in southern England. *Sociologia Ruralis*, 27, 103–22.

Zelinsky, W. 1970: Beyond the exponentials: the role of geography in the great transition. *Economic Geography*, 46, 499–535.

2

The Changing Organization of the Global Economy[1]

Peter Dicken

> The world economy is changing in fundamental ways. The changes add up to a basic transition, a structural shift in international markets and in the production base of advanced countries. It will change how production is organized, where it occurs, and who plays what role in the process.
>
> Cohen and Zysman, 1987:79

> The talk today is of the 'changing world economy'. I wish to argue that the world economy is not 'changing'; it has *already changed* – in its foundations and in its structure – and in all probability the change is irreversible.
>
> Drucker, 1986:768

There is widespread acceptance that something fundamental is happening, or indeed has already happened, in the way the global economy is organized. However, precisely what these changes are, and how they have been caused, is a matter for considerable debate. The term 'globalization' has become common currency, yet it is a currency which is already in danger of being overvalued. The world, we are often led to believe, is becoming increasingly homogenized economically, and perhaps culturally too. Phrases such as 'global market-place' and 'global factory' have been added to the older McLuhanism of the 'global village'. But there are dangers in the too-ready acceptance of such seductive terms. There is no doubt that profound changes have been, and are, occurring in the world. But the label 'globalization' is too often applied very loosely and indiscriminately to imply a totally pervasive set of forces and changes with all-embracing effects on countries, regions and localities. We need to adopt a more discerning and less simplistic approach in articulating the nature and processes of globalization and in assessing its implications.

The aim of this chapter is to explore, in a brief and inevitably superficial manner, both the surface manifestations of changes in the organization of the global economy and the major underlying processes of change. Three themes are addressed. First, some indicators of the changing organization of the global economy are described very briefly. Second, the major causes of such change are identified as the rearticulation and reorganization of the production chain, which is resulting from the complex interplay between the *internationalization of capital* (primarily the behaviour of *transnational corporations*) and the actions of national governments, set within the volatile context of pervasive *technological change*. Third, I suggest at least some elements of a geographical research agenda.

CHANGES ON THE SURFACE OF THE GLOBAL ECONOMIC MAP

There is a rich, though often indigestible, menu of data which describes the broad features of global economic change using nation-states as the basic statistical unit. These are dealt with in detail elsewhere (Dicken, 1992b). Here, I simply summarize some of the major points:

(1) *Production*: There has been a considerable global redistribution of industrial production during the past 30 years in particular, both among the industrialized economies and also towards some developing market economies. But such redistribution is far more limited than is often suggested. Within the group of industrialized countries the two major developments have been the relative decline of the United States as an industrial producer and the spectacular growth of Japan. Within the developing economies only a very small number of countries have obtained a manufacturing production base of any significance. So although industrial production is no longer exclusively a 'core' activity, as in the old international division of labour, it remains highly concentrated geographically in global terms.

(2) *Trade*: One of the most striking features of the global economy in the last few decades is that international trade has grown much more rapidly than international production – a clear indication of increased international *integration* in the global economy. For virtually the whole of the post-war period the major driving force in the growth of international

trade has been the manufacturing sector. Only recently has the growth of international trade in services (notably commercial, and especially financial, services) taken on a major propulsive role; a manifestation of the increasing integration of manufacturing and services in the global economy (GATT, 1989).

In terms of geographical origins, trade is less concentrated than production. But it is still dominated by the industrialized economies: 80 per cent of all world manufactured exports are generated by the industrialized countries and around half of world trade is trade *within* this group. However, a small number of newly industrializing countries (NICs) have emerged as global players: the East and South East Asian NICs in particular. Their annual growth rates have continued to be very high indeed. Geographically, international trade exhibits a number of features, notably: strong regional patterns (especially in Europe but also in North America); the heavy dependence of NICs on the United States market; and the particularly sensitive relationship between the United States and the European Community on the one hand and Japan on the other.

As far as both global production and trade are concerned it is clear that we have moved strongly towards a *multi-polar* structure in which three major regional blocs are evident: North America, the European Community, and East and South East Asia (focused on Japan). This 'triad', to use Ohmae's (1985) term, sits astride the global economy like a modern three-legged Colossus. These three regions dominate global production and trade: 77 per cent of world exports is generated by them; 62 per cent of world manufacturing output is produced within them. They are without doubt the 'mega-markets' of today's global economy.

(3) *International direct investment*: The growth of international direct investment has been particularly rapid in the post-war period, even more rapid than that of international trade. There was some slackening of investment growth between 1974 and 1983 (OECD, 1987) but then a very marked reacceleration during the 1980s. Indeed, it has been calculated that foreign direct investment (FDI) grew four times as fast as gross national product (GNP) during the 1980s, compared with twice the rate of GNP growth in the 1960s (Julius, 1990). During this period, the geographical origins of direct investment have diversified. Although the United States and the United Kingdom are still the dominant sources (roughly half the world total), there has been rapid growth in outward investment from most western European countries and, especially, from Japan. In 1960 Japan accounted for less than 1 per cent of the world

direct investment stock; by 1985 its share had increased to 12 per cent. In terms of annual direct investment *flows*, Japan is now probably the most important source nation. There are also significant indications of substantial foreign direct investment emanating from some of the leading NICs.

The overwhelming bulk of direct investment is placed *within* the developed market economies themselves. Only one quarter of the total is located in developing countries, compared with two-thirds in 1938, and this investment is extremely uneven geographically. A particularly significant development has been the blurring of the formerly clear distinction between those countries which were primarily home countries (sources) of FDI and those which were primarily host countries (destinations). The predominant pattern today is one of complex *cross-investment* among the industrialized economies and a generally closer balance between inward and outward investment flows (OECD, 1987).

Most significant of all has been the dramatic change in the position of the United States as a host country for foreign direct investment. For every leading investing country the United States has become the major destination. Its direct investment position has been transformed from one of being predominantly a home country for FDI to one in which the ratio of outward to inward investment is virtually in balance. The same certainly cannot be said of Japan. While Japanese outward investment has grown rapidly, there has been limited growth of inward investment (a ratio of roughly 12:1). In sectoral terms the most significant recent development has been the massive increase in foreign direct investment in services, particularly circulation and business services (UNCTC, 1988). Although some of these services are tradeable they frequently either require or benefit from close geographical proximity between seller and buyer. The internationalization of financial services has become especially marked, as Thrift (1989) has demonstrated.

PROBING BENEATH THE SURFACE: IDENTIFYING THE MAJOR FORCES OF CHANGE IN THE ORGANIZATION OF THE GLOBAL ECONOMY

Such international data on production, trade and direct investment in the global economy are important. They point to *some* of the major organizational changes that are occurring and they are certainly relevant to national policy-makers. But the use of 'national boxes' as the units for data collection obscures more than it reveals. This is not to claim (as some indeed do) that the nation-state is no longer a significant actor in the global economy – I will argue that it is – but that the processes of change cut across national boundaries in extremely complex ways. Thus, national production data tell us nothing about the actual structure of the production system; national trade data do not reveal the increasing tendency for a great deal of trade to occur *within* individual companies as intra-firm trade; national FDI data fail to capture the increasing diversity in the modes of international involvement by transnational corporations.

Historically, of course, there was a very close relationship between the organization of the entire production process and nationally bounded units. As Hobsbawm (1979) observed, the production process was 'primarily organized *within* national economies or parts of them. International trade ... developed primarily as an exchange of raw materials and foodstuffs ... [with] products manufactured and finished in single national economies ... In terms of production, plant, firm and industry were essentially national phenomena' (1979:313). The severing of this relationship between national economic space and the production process (and especially the national regulation of that process) is one of the symptoms identified in Lash and Urry's (1987) study of 'disorganized' capitalism. But what exactly has caused the transformation of a formerly state-centred economic system to a global economic system?

One way of getting a handle on the processes of change is to begin with a conceptualization of the basic *production chain* (alternative terminology includes value-added chain, value chain, filiere). Figure 2.1 presents a very simplified picture of the basic production chain and of a system of linked production chains. Its core is the transactionally linked sequence of functions in the centre of the diagram, which can be applied to all segments of the economy. Common to the entire chain of functions are the technological processes involved in production itself and in the physical movement of the constituent elements. Although they are identified separately in the diagram, it is not implied that they are in any way exogenous; on the contrary, they are intrinsic elements of the system. Production chains and production systems have to be coordinated, controlled and regulated, and they also have to be financed. Both of these processes apply to the system as a whole, but they may be either endogenous or exogenous to it.

The production chain and its associated processes may be structured in a variety of different ways. Two dimensions are particularly important (Porter, 1986). One is explicitly *geographical*. Functions may be geographically concentrated or geographically dispersed. The major development of the past few decades is that the potential geographical scale has become extended to the global level. The other dimension is

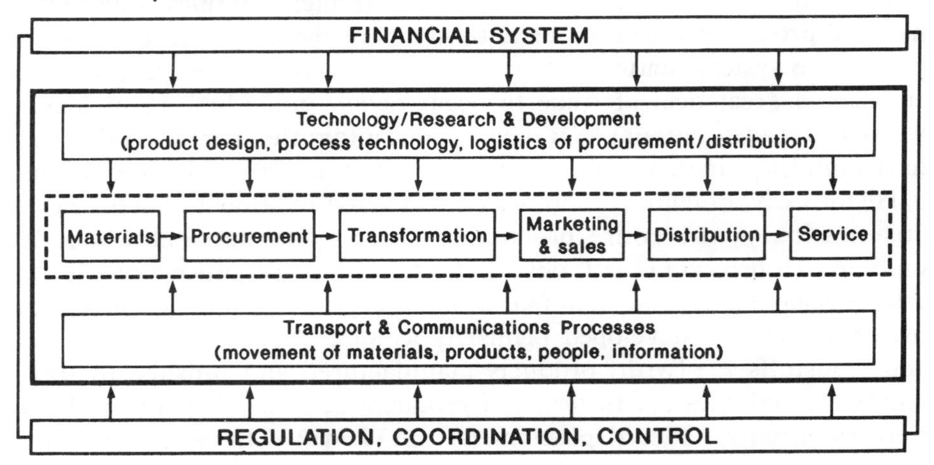

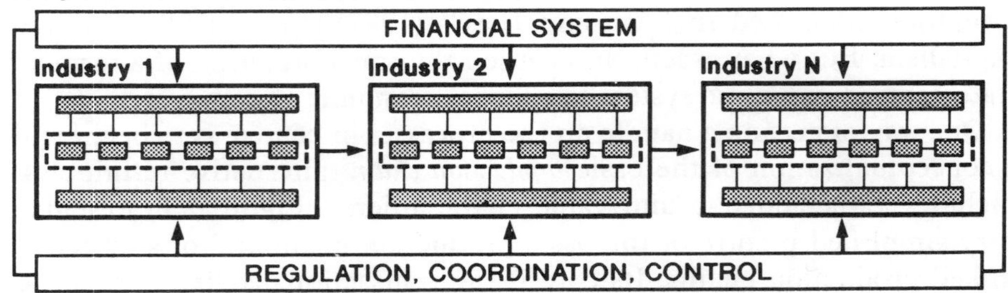

Figure 2.1 The basic production chain
(*Source*: Dicken, 1992b, Figure 7.1)

organizational; the degree of the tightness or looseness of control or coordination in a particular production system. Following the work of Coase (1937), Williamson (1975) and Teece (1980), coordination may be implemented either via 'the market', through the externalization of transactions between independent firms, or within a particular organization or firm, through its internalization of transactions. In fact these two extremes of 'markets' and 'hierarchies' do not encompass the full range of possibilities. There are other forms of inter-unit coordination, such as subcontracting relationships or collaborative ventures, which are increasingly significant components of the global economy.

In today's global economy there is no doubt that the pre-eminent form of economic organization is the transnational corporation (TNC). But we need to get away from the simplistic notion that a TNC's activities are defined only in terms of *ownership* of facilities and that the only valid measure of such activities is foreign direct investment. The most recent United Nations survey of TNCs in world development (UNCTC, 1988) adopts a much broader definition of TNCs than that used in its earlier surveys, and recognizes that it need not necessarily imply equity investment. In fact, Cowling and Sugden suggest a definition of the TNC which makes it possible to incorporate a wide range of different modes of international involvement: 'a transnational is the means of coordinating production from one centre of strategic decision making when this coordination takes a firm across national boundaries' (Cowling and Sugden, 1987:60).

Foreign direct investment is one form which such strategic coordination may take. It is a mechanism for locating all or some parts of the production chain outside the firm's home country. The precise locational orientation of FDI can be broadly classified into market-oriented or supply/cost-oriented activity, including the establishment of export-platform and offshore sourcing plants. Such foreign *direct* investment has been extremely important in the industrialization of some NICs (such as Singapore) but far less significant in others (such as South Korea). TNCs have the capacity to locate and relocate some or all of the production chain functions at an international or global scale. There is, indeed, *some* geographical symmetry between the internal division of labour within TNCs and the broader international division of labour between countries and regions. But, as Walker (1989) correctly observes, it is far more complex than originally suggested by Stephen Hymer (1972) and by much of the global city literature. It is also changing over time as, for example, the major TNCs differentially relocate specific parts of the control functions or their research and development (R & D) functions (Howells, 1990).

Shifting geographical patterns of FDI are one significant indicator of transnational corporate activity in the global economy. But FDI statistics give no indication of the increasing diversity in the forms of international corporate involvement. Many of these do not involve ownership and equity relationships but are rather various forms of collaboration between legally independent firms in different countries. Two modes of inter-firm collaboration are especially significant: global strategic alliances and international subcontracting.

Strategic alliances between firms across national boundaries are not new in themselves (Kindleberger, 1988). What is new is their current scale, their proliferation, their increasingly complex multilateral nature and the fact that they have become *central* to the global strategies of many firms rather than peripheral to them. Companies are increasingly forming not just single alliances but *networks* of alliances. The position is one of polygamy rather than monogamy: 'few companies have only a single alliance. Instead, they form a series of alliances, each with partners that have their own web of collaborative arrangements. Many companies are at the hub of what are often overlapping alliance networks which frequently include a number of fierce competitors' (Business International, 1987:113–14). Strategic alliances are especially prominent in certain sectors, notably automobiles, telecommunications, electronics, aerospace and pharmaceuticals, all of which are being transformed by the proliferating spiders' webs of collaborative ventures. There are no data comparable to those for FDI which measure comprehensively the incidence of strategic collaborations; they have to be analysed on a case-by-case basis (for a recent example, see Wells and Cooke, 1991).

Rather more attention has been paid by geographers to a second form of inter-firm collaboration: the contractual relationships between firms and their component suppliers. Much of the empirical research in this area has been in just a few sectors, most notably automobiles (Holmes, 1986; Sheard, 1983). However, international subcontracting has been one of the most important *indirect* ways in which foreign firms have been heavily involved in the industrialization of the NICs. In a pioneering paper published nearly twenty years ago, Hone (1974) argued that it was *the* major force in the development of export-based industrialization in East and South East Asia; far more important than foreign *direct* investment. Certainly in South Korea, where direct entry of foreign capital has been carefully restricted, a major factor in the growth of domestic firms of all sizes, including the giant *chaebol* (conglomerates), has been their function as subcontractors to foreign companies.

During the 1960s, the geographical scale at which subcontracting networks operate shifted from highly localized networks (the traditional industrial districts of most older manufacturing cities) to nationally and internationally extensive networks, based on what Sayer (1986) – following Schoenberger (1982) – called 'just-in-case' procurement systems. During the 1980s there has been something of a scale reversal as a result of two developments: the spread of 'just-in-time' procurement systems and the re-emergence of new industrial districts based upon flexible systems of production (Piore and Sabel, 1984; Scott, 1988; Amin and Robins, 1990; Lovering, 1990).

Conventionally, much subcontracting has involved a firm putting out certain aspects of its operations to specialist suppliers while continuing to engage in manufacturing itself. Recently, however, a rather different organizational form has emerged: the *vertically disaggregated network* organization in which almost *all* the functions in the production chain, other than those of central control and coordination, are contracted to independent firms but in which the final product is marketed under the parent company's brand name. Benetton is often used to exemplify this tendency but rather better examples at the global scale are NIKE (see Donaghu and Barff, 1990), Reebok and Amstrad.

What these various modes of international organizational involvement reflect is increasing complexity and flexibility in the articulation of the production chain and in the positioning of functions along the coordination/control and geographical dimensions. To some writers, the various forms of strategic alliance and subcontracting networks are an indication of 'the distinction between Fordist multinational organizational forms and those of post-Fordist globalized organizational forms. The globalized firm has indistinct and shifting boundaries, we may expect it to be networked or distributed in organizational structure rather than being hierarchical, and it may penetrate and exploit space by proxy or in cohort with other firms rather than in "isolation"' (Wells and Cooke, 1991:17). It would be a mistake to assume, however, that all TNCs are proceeding down a single path. What we have in reality is a spectrum of organizational forms, from the conventional TNC controlling its hierarchically organized, wholly owned overseas affiliates from a centralized headquarters, to the flatter network structures, which may be equally tightly controlled and coordinated, although not through ownership. Variety rather than homogeneity is the rule.

Such variety in organizational form is related to specific developments in product and process technology and in the space-shrinking technologies. It is now generally accepted that we are in the process of transformation to a new techno-economic paradigm (Freeman, 1987; Dosi et al., 1988) based upon information technology. Three sets of change are especially significant (Perez, 1985):

1 *increasing information intensity* in the production process and reduced materials and energy intensity, which is leading to the redesign of many products;
2 *much-enhanced flexibility of production*, which is breaking down the traditional economies-of-scale relationships and permitting much smaller volumes of a greater variety of products and product variants;
3 *reduction in the volume of labour* needed in the production process together with *changes in the type of labour* required (skill changes).

Clearly, the potential of the new flexible technologies is immense and their implications are enormous for the nature and organization of economic activity on all geographical scales. But precisely what these implications are is the subject of heated debate (see, for example, Gertler, 1988; Schoenberger, 1988a, 1989; Scott, 1988; Coombs and Jones, 1989; Sayer, 1989; Amin and Robins, 1990; Lovening, 1990). The post-Fordist writers assert that flexible production (and flexible organization) has now virtually displaced the formerly hegemonic Fordist system with a new production paradigm. This is reflected not only in some of the forms of networked organizations referred to earlier but also in the newly discovered industrial districts of middle Italy, Silicon Valley and the like. However, not all agree with this diagnosis. Coombs and Jones, for example, argue that 'it is premature to diagnose a global trend toward a unique and well-defined successor to Fordism as a paradigm for production organization' (1989:115). Amin and Robins make a similar point. They observe that the 'new orthodoxy' takes

> one element of the present restructuring process, crystallized in the concept of flexibility, and . . . [projects] it forwards as the one guiding principle of a new social era.. Its conceptual framework is structured around a simplistic binary opposition [mass-flexible] . . . socio-historical processes are more complex and contradictory . . . The new is not marked by an absolute and fundamental break from the old: the old order of things, does not, cannot, simply and conveniently disappear. (1990:25)

A similar point can be made about developments in the strategies of TNCs which, we are often led to believe, are adopting pure globalization *strategies* of both production and marketing: global production for a global market. In fact, the intensification of *global* competition in a world which still retains a high degree of national, regional and *local differentiation* creates a whole set of internal tensions between globalization strategies on the one hand and localization strategies on the other. International firms are increasingly finding themselves trapped between two apparently counterpoised forces. One is the pressure to *integrate the firm's operations on a global basis*; the other is the pressure to *respond to national or local differences* and to tailor or customize activities to fit such variations. Some argue that it is possible to categorize industries into two discrete types: global industries on the one hand and multi-domestic or nationally responsive industries on the other. Each has very different attributes. One conventional wisdom is that more and more industries are becoming transformed into global industries in which functional integration occurs across national boundaries in terms of R & D, production and marketing activities. It is arguable as to whether whole industries can be so simply categorized; but even if this is so, it does not necessarily follow that all firms in a particular industry will follow the same strategy. Some of the recent work on competitive strategies strongly rejects the polarization of strategic choice between global and local orientations. In a world of 'a much more complex set of environmental forces ... firms must respond simultaneously to diverse and often conflicting strategic needs ... In the emerging international environment ... there are fewer examples of industries that are pure global, textbook multinational, or classic international. Instead, more and more businesses are being driven by *simultaneous* demands for global efficiency, national responsiveness and worldwide learning' (Bartlett and Ghoshal, 1987:10, 12).

The changing global organization of the production system is a complex interweaving of diverse tendencies. Although there is undoubtedly increasing *global integration* of the transactionally-linked sequence of functions in the production chain, there remains a high degree of local differentiation. The same may be said of developments in the financial system, without which the global production system could not operate. Drucker (1986) has argued that the 'real economy' of flows of goods and services and the 'symbol economy' of capital movements, exchange rates and credit flows seem increasingly to be operating independently of each other. This view is consistent with Thrift's conceptualization of a new international financial system, which, he argues, has become in effect a separate state, largely outside the regulatory powers of the nation-state. 'We are left', he suggests, 'with a strange world, one in which money capital flows freely, and is becoming less and less regulated, while movement of goods in the "productive" economy has become more and more negotiated and regulated. Yet this productive economy is increasingly susceptible, in principle if not in practice, to the institutions of money capital' (Thrift, 1990:1136). There is much truth in this, but there still remains a considerable variety of nationally regulated financial systems within an uneven, and variable, pattern of deregulation.

This leads to some observations about the *role of the state* in the changing organization of the global economy. Historically, the state was the primary regulator of its national economic system; the world-economy could quite legitimately be conceptualized as a set of interacting national economies (Radice, 1989). This is, of course, no longer the case. Not only has the degree of interdependence and interconnection within the world-economy increased, but also the emergence of the TNC has produced a major rival to the state's traditional role as the dominant global economic institution. These developments have led some to pronounce the virtual demise of the state as an important force in the world economy (see, for example, Kindleberger, 1969:207; Johnston, 1982: 61). Such assertions greatly distort and oversimplify the real position.

It is certainly true that an individual state's degrees of economic freedom – its economic autonomy – are constrained by the actions of both other states and, especially, TNCs. Nevertheless, national governments, whether singly or collectively, continue to play a most significant role in the global economy. Although the processes of competition no longer 'fracture along national lines' (Radice, 1989), national boundaries still create significant differentials on the global economic surface. (If this were not the case there would be no need for the current efforts to create, by legislation, a single European market.) National political spaces are still among the most important ways in which location-specific factors of both supply and demand are 'packaged'. Political boundaries create discontinuities of varying magnitudes in the flows of economic activities. States can modify, create or destroy comparative advantage. There is no doubt, for example, that the state has played a central role in the industrialization of all the NICs, though in differentially specific ways.

States operate within a world system of differential and unequal power relationships both with other states and with TNCs. In an increasingly integrated and interdependent global economy, states are locked into intensive economic competition with other states: to enhance their international trading position and to capture as large a share as possible of the gains from trade; to create and attract productive investment which, in turn, enhances their competitive status (for a discussion of the 'competitive advantage of nations', see Porter, 1990). At the same time, states are locked into a complex interaction with TNCs. There is a fundamental conflict of interest between states and TNCs, particularly globally integrated TNCs. The global corporation understandably seeks to maximize its freedom to locate its functions in the most advantageous locations for the corporation as a whole in the pursuit of global profits. Equally understandably, an individual national government prefers to have the entire production chain, or at least the highest value-adding functions, within its own boundaries in order to maximize positive spin-off effects. In some circumstances, a state may well have the power either to encourage or to inhibit the globally integrated policies of TNCs (Doz, 1986). Hence, the relationship between the two dominant sets of institutions in the global economy is inevitably an uneasy one, which tends to fluctuate between conflict and cooperation and which may be very variable between different sectors of the economy.

In a highly integrated global system there are few government economic policies, whether macro- or micro-, which do not have an international dimension. Two policy areas are currently the subject of considerable international (and national) controversy and are directly relevant to my argument: trade regulation, and host-country attitudes to foreign firms. Both policy areas are deeply imbricated with the current trend towards the strengthening of regional economic blocs.

Within the auspices of the GATT (General Agreement on Tariffs and Trade), the principle underlying international trade after World War II was multilateral agreement and the application of the 'most favoured nation' principle. Although the GATT has clauses which permit deviations from these basic principles, the circumstances are clearly defined and the GATT is the ultimate arbiter of disputes. Until the mid-1970s, a major feature of the international trading environment of the post-war years had been the progressive liberalization of trade and, especially, the decline in tariff barriers through the various GATT rounds. For the industrialized countries, the average tariff on manufactured goods fell from 40 per cent in 1947 to between 6 and 8 per cent in 1973. As a result of the Tokyo round of the GATT, tariffs on manufactured goods averaged 6 per cent in the European Community, 5.4 per cent in Japan and 4.9 per cent in the United States by the early 1980s. In fact, Japan's tariff cuts went further than this (World Bank, 1987).

However, since 1974 there has been a clear and marked increase in the use of non-tariff barriers (NTBs). Most recently these have become part of a more systematic tendency towards the adoption of 'strategic' trade policies in a number of countries, notably the United States. Bhagwati (1988) distinguishes between two types of NTB:

1 *'high track' restraints*, which bypass the GATT rule of law. These are the 'visibly and politically negotiated' restraints negotiated by trading partners: the voluntary export restraints (VERs) and orderly marketing arrangements (OMAs) which have proliferated in a variety of industries including textiles and clothing, automobiles, consumer electronics, steel, footwear, machine tools, etc. The OECD estimates that such NTBs today affect one quarter of all industrialized country imports. Between 1973 and 1983, the proportion of world automobile trade covered by such restrictions increased from less than 1 per cent to almost 50 per cent. Kostecki (1987) identified 113 major VER

agreements. These figures *exclude* the most comprehensive case, the Multi-Fibre Arrangement. Of the 113 cases, almost 90 per cent protected the EEC, USA and Canadian markets. Almost two-thirds involved exports from Asia, primarily Japan and South Korea. Overall, Kostecki estimated that some 12 per cent of world non-fuel trade in the mid–1980s was covered by export-restraint arrangements. A third estimate, by the World Bank, showed that 16 per cent of industrial countries' imports from other industrialized countries and 21 per cent of imports from developing countries were subject to 'hard core' NTBs in 1986, compared with 13 per cent and 19 per cent respectively in 1981;

2 *'low track' restraints* which 'capture' and 'pervert' the GATT rules. Such measures include anti-dumping provisions and countervailing duties. These play 'legitimate roles in a free trade regime ... but not if they are captured and misused as protectionist instruments' (Bhagwati, 1988:48). There has been a surge in *anti-dumping* measures implemented by the USA, EC, Canada and Australia in particular. In 1980, the USA initiated 22 anti-dumping actions; in 1985, 65. The comparable figures for the EC for the same years were 25 and 42.

Such protectionist measures are being set within a more general structure of trade legislation put in place by individual countries (and groups of countries). The most significant example is the new trade legislation in the United States, the 1988 Omnibus Trade and Competitiveness Act – what Bhagwati calls the 'ominous' trade and competitiveness act. This elevates the principle of unilateralism over multilateralism, and incorporates a 'crowbar' in the 'super 301' clause which aims to achieve reciprocal access to what the USA defines as unfairly restricted markets. Significantly, this clause is directed to entire *countries*, not individual sectors. In its first application in May 1989, three countries – Japan, India and Brazil – were named as 'priority countries'; other were 'warned' that they were being watched. In a related set of measures, the USA has instigated a 'Structural Impediments Initiative' in the attempt to open up what are seen to be structural restrictions on access to the Japanese market.

A second major development of great significance is the strengthening of regional economic groupings in the west: notably the completion of the single European market in 1992 and the US-Canada Free Trade Agreement. In terms of the changing competitive environment, one of the biggest developments facing both national governments and business enterprises is undoubtedly the completion of a single European market. The stimulus to complete the integration of the EEC was the fear that

Europe was losing out in the global competitiveness stakes and that the continuing fragmentation of the community was a major source of competitive weakness for both European firms and member-states. In some respects, whether or not immediate integration actually happens is less significant than the fact that the Single European Act has set in motion a whole chain of developments, not the least being the urgent evaluation of competitive strategies by firms in all industries. As Callingaert (1988:37–8) observes, '1992 is a process not an event . . . The process will continue beyond 1992. Indeed, it is unlikely that a specific point will be indentifiable at which the internal market becomes "complete".'

For firms and governments outside the EC, the key issue is not so much the removal of internal barriers *per se* but rather the fact that the removal of such barriers will be set within a single *external* community boundary. For firms not established in Europe but anxious to maintain or achieve market access, the overriding concern is the possible emergence of a *fortress Europe*. Certainly, the various national bilateral trade agreements will be replaced by a single community arrangement. They will be far more contentious in some sectors than others, as the experience of the automobile industry shows (Dicken, 1992a). At present, most of the controversy surrounds those industries in which Japanese market penetration is very high (automobiles, electronics). In so far as, in general, the direct Japanese presence in Europe is still very small, it is inevitable that the level of Japanese FDI in Europe will grow substantially and possibly take on more diverse forms than hitherto. There are signs, too, of increasing direct investment in Europe by South Korean and Taiwanese firms, particularly in electronics.

A further complication has arisen with the remarkably rapid opening up of the Eastern European economies following the political changes of 1989/90. East Germany has already been united with West Germany to form the biggest European economy. The former Soviet satellites, especially Czechoslovakia, Hungary and Poland, are seeking to incorporate market forces and to attract foreign investment. For the NICs in particular, an important question is whether at least some labour-intensive manufacturing activity initiated by US and European firms may be diverted to Eastern Europe.

The late 1980s also saw a strengthening of a North American regional trading bloc. The US-Canada Free Trade Agreement was signed in 1988. Its objective is to remove bilateral tariffs between the two countries over a 10-year period and to remove other trade barriers. In 1990, a further dimension was added to the United States' recent conversion to policies of regional economic integration with the announcement of the start of negotiations with Mexico on a possible free-trade agreement between the two countries. Further, but more tentatively, President Bush has suggested that a longer-term objective could be the incorporation of the whole of Latin America into an Americas free-trade area stretching from Anchorage to Tierra del Fuego.

The emergence of explicitly 'strategic' trade policies is leading to more direct linkage with a second key policy area: the treatment of foreign direct investment. Generally speaking, the developed market economies have adopted a liberal attitude towards FDI, a reflection of the fact that they are the major source countries. There are, of course, exceptions to such a generalization. During the 1980s, however, substantial changes occurred and restrictive policies were eased in almost all countries, including many developing countries (UNCTC, 1988). Conversely, however, attitudes in the United States towards the massive upsurge in inward investment hardened substantially, even though, so far, there have been no major policy shifts. But there is now intense debate – and a good deal of acrimony – about the alleged foreign takeover of the USA (Tolchin and Tolchin, 1988; Glickman and Woodward, 1989).

Within the array of possible policy measures adopted by countries towards inward investment, there are two which are increasingly pervasive but somewhat contradictory. The first is the set of policies which can be bundled together under the heading of *performance requirements*. In the past, attempts to impose such requirements on foreign firms were predominantly made in developing countries; more recently they have become increasingly common in developed countries. Three types of requirement have become most apparent:

1 the requirement that a specified proportion of a plant's inputs should be sourced locally. This *local content* stipulation has become especially common in assembly industries such as automobiles and electronics, particularly in Europe, and is the focus of a great deal of controversy both between national governments and private investors and also between national governments themselves. Recently it has become enmeshed within the anti-dumping procedures referred to earlier;
2 insistence on a minimum level of exports from the host country plant;
3 a requirement relating to technology transfer, including attempts to persuade firms to set up R & D facilities in the host country.

The second set of policies towards FDI runs somewhat counter to the essentially restrictive nature of performance requirements: the *competitive bidding* which takes place between nations (and within nations) for internationally mobile investment. Again, this is not a new development. But there is little doubt that it has intensified very markedly indeed. One result of intensified competition for internationally mobile investment has been both a general escalation of incentive levels and also efforts to 'differentiate' the national offering, through a whole battery of financial and non-financial incentives and through aggressive promotional activity. The problem is, of course, that the upward escalation of incentives may not actually increase a country's share of the investment market if its competitors behave likewise.

Trade and FDI policies exert an extremely important influence on the geographical configuration of the production chain and, hence, on the organization of the global economy. The application of non-tariff barriers and of firm performance requirements, either singly or collectively, has undoubtedly influenced the geography of TNCs' investment decisions, as the Japanese experience shows very clearly (Dicken, 1988). Just as, in an earlier era, national tariff barriers encouraged foreign firms to jump over them to establish a direct presence in a protected market, so NTBs and local content requirements have a similar effect. In some instances, both trade and FDI policies have been applied differentially by economic sector. In most national economies, certain sectors are heavily protected not just from import competition but also from the actual entry and participation of foreign companies. Recently there has been a clear, but uneven, trend towards the opening up or *deregulation* of some significant sectors, notably financial services and telecommunications, both of which are crucial components of the overall production system (see Figure 2.1). However, the more ambitious attempts at the global scale to remove NTBs, to open up financial, telecommunications and other service sectors and to introduce 'trade-related investment measures' within the GATT have foundered with the collapse of the Uruguay Round of negotiations. What the extended and often acrimonious GATT negotiations have demonstrated, however, is the continuing significance of the state in the global economy.

CONCLUSION: TOWARDS ELEMENTS OF A RESEARCH AGENDA

I began this chapter with a plea for a more discerning and less simplistic approach to understanding global economic change. We live in an extremely complex world yet, too often, we try to deal with it using highly simplistic concepts and 'easy fixes'. My major point is that the contemporary global economy is one which is no longer state-centred; but that states are not insignificant actors, mere cyphers in a world in which global TNCs slice their way at will through national boundaries and render all state policy-makers impotent. Reality is a complex interplay between these two major sets of global actors, a nexus of intricate power relationships and bargaining systems, set in a volatile technological environment.

The potential research agenda is clearly enormous. I want to focus on just three issues, making a plea, in particular, for greater collaboration between researchers, who presently tend to operate within hermetically sealed containers.

First, there is a major need for more penetrating conceptual and empirical research on the changing technical and organizational configuration of the production chain or filiere and on its evolving geographical configuration at different spatial scales. The production chain *is* the fundamental building block. How it is regulated, coordinated and controlled is of the utmost importance to variations in economic and social well-being throughout the world. Walker (1989) is right to emphasize the centrality of production and to urge so-called 'corporate geographers' or 'geographers of enterprise' to recognize this fact. He is wrong, however, in inferring that none does so. He is even more wrong, in his desire to compose his 'requiem for corporate geography', to dismiss the organization as a legitimate and necessary object of research. What is needed is a greater degree of interaction and integration between those who research the production system *per se* and those who focus on the organization. The way in which transactions in the production chain are organized has come to be recognized as a key issue. But we need to beware of accepting uncritically such simple dichotomies of transactional organization as markets and hierarchies, and to explore the much greater variety of possible organizational configurations between the elements of the production chain. In so doing we should be able to move beyond the highly simplistic – but, in its day, significant – conceptual parallel drawn by Hymer (1972) between the internal division of labour of large TNCs and the international division of labour between countries and regions.

We need, as well, to take *all* elements of the production chain seriously rather than, as at present, ignoring most of what happens beyond the transformation stage. In particular, there has been a remarkable neglect of marketing functions in the spatial literature. Such geography of marketing as does exist is, in practice, confined to the isolated island of the geography of retailing. Wells and Cooke (1991) rightly observe that 'much more research needs to be conducted on the geography of markets in relation to corporate strategy' (p. 102) and argue for clearer links to be made between 'the organization of firms in terms of markets, and the organization in terms of labour and production' (p. 102). A major weakness of the 'new international division of labour school' was precisely its neglect of markets in the shift of production towards developing countries (Elson, 1988; Schoenberger, 1988b). Lastly, there is a need for empirical studies to move beyond the very narrow range of sectors currently favoured. Thrift has rightly called for greater attention to be given to the international financial system, but there are many other neglected sectors. One is the food industry, so central to all societies and yet an industry which tends to be almost totally neglected outside the subdiscipline of 'agricultural geography'. Again, the need for the breaking down of academic barriers is apparent.

A second major need is to fill one of the major lacunae in both economic and political geography research: international trade. Such trade is the primary manifestation of increasing integration in, and changing organization of, the global economy. It is, as we are currently witnessing, a major focus of geopolitical tension. It must therefore occupy a key position in our research agenda. This is not a plea for more studies of geographical patterns of international trade or of the modelling of trade flows, useful as such studies can be. As Johnston (1989:346) has pointed out, a problem with such studies is that they lack a satisfactory basis in explanatory theory. He is right to argue that 'trade cannot be studied in isolation; the part cannot be appreciated independently of an understanding of the whole'. But there is considerable doubt about whether the theoretical framework he favours as the 'best available' is as robust as he suggests, given its basis in world systems analysis and dependency theory, both of which are subject to substantial criticism (Sklair, 1991).

Any theoretical framework must incorporate the two global actors I have focused on in this chapter: TNCs and states. First, geographical studies of international trade must be explicitly related to the changing configurations of the production chain. A large and growing proportion of world trade consists of transactions within and between TNCs; we need to learn a great deal more about this. Second, we need to pay specific attention to trade policies. Yet where is the geographical research on the issues involved in the GATT Uruguay Round, on the growth of managed trade and of the emergence of strategic trade policy? Neither economic nor political geographers adequately address these issues. Perhaps economic geographers regard them as the political geographers' territory. But the revitalized 'new' political geography based on world systems analysis is largely silent on these most critical politico-economic processes. Taylor (1989) has a brief section on trade policy and the world-economy but is overly constrained by his Wallersteinian framework.

Agnew and Corbridge (1989) incorporate a brief discussion of global trade relations into their analysis of the dynamics of geopolitical disorder. But the leading political geography journal, *Political Geography*, has published little or nothing on the topic, a reflection, presumably, of the lack of interest by political geographers in conducting such research. There is a very real need for the development of collaborative research by economic and political geographers.

My third, and final, agenda item is in a similar collaborative vein. We need to develop research on the *interactions* between TNCs and states which, I have argued, are central to the changing organization of the global economy. One possible direction is to explore the nature and outcomes of the bargaining processes in which TNCs and states are engaged, such as investment projects and issues relating to the environment. This is an area of research which is generally underdeveloped, and not only by geographers. In the non-geographical literature, most attention has been devoted to the natural resource sectors in developing countries, and it has been based on the notion of the 'obsolescing bargain' (Kobrin, 1987). Clearly, this is a difficult research area, not least because bargaining negotiations tend to be conducted in some secrecy. But, as studies in the vehicle industry have demonstrated (for example, Seidler, 1976), it is by no means impossible to reconstruct particular cases and to identify significant geographical outcomes.

There are many other possible agenda items. In selecting just three I have merely tried to suggest some of the kinds of research which I believe are needed for increasing our understanding of the changing organization of the global economy. The most important point, however, is not so much the specific research topics themselves but rather the need for breaking down subdisciplinary barriers within geography. By the very nature of our discipline, geographers have a unique opportunity to contribute towards a more holistic understanding of global economic change.

NOTES

[1] This paper draws extensively on Dicken, 1992.

REFERENCES

Agnew, J. and Corbridge, S. 1989: The new geopolitics: the dynamics of geopolitical disorder. In R. J. Johnston and P. J. Taylor (eds), *A World in Crisis?: geographical perspectives* (second edition), Oxford: Basil Blackwell, chapter 10.

Amin, A. and Robins, K. 1990: The re-emergence of regional economies? the mythical geography of flexible accumulation. *Environment and Planning D: Society and Space*, 8, 7–34.

Bartlett, C. A. and Ghoshal, S. 1987: Managing across borders: new strategic requirements. *Sloan Management Review*, 7, 7–17.

Bhagwati, J. N. 1988: *Protectionism*. Cambridge, MA: The MIT Press.

Business International 1987: *Competitive Alliances: how to succeed at cross-regional collaboration*. New York: Business International Corporation.

Callingaert, M. 1988: *The 1992 Challenge from Europe: development of the European Community's internal market*. Washington, DC: National Planning Association.

Coase, R. 1937: The nature of the firm. *Economica*, 386–405.

Cohen, S. S. and Zysman, J. 1987: *Manufacturing Matters: the myth of the post-industrial economy*. New York: Basic Books.

Coombs, R. and Jones, B. 1989: Alternative succesors to Fordism. In H. Ernste and C. Jaeger (eds), *Information Society and Spatial Structure*, London: Belhaven Press, chapter 8.

Cowling, K. and Sugden, R. 1987: Market exchange and the concept of a transnational corporation. *British Review of Economic Issues*, 9, 57–68.

Dicken, P. 1988: The changing geography of Japanese foreign direct investment in manufacturing industry: a global perspective. *Environment and Planning A*, 20, 633–53.

Dicken, P. 1992a: Europe 1992 and strategic change in the international automobile industry. *Environment and Planning A*, 24, 11–32.

Dicken, P. 1992b: *Global Shift: the internationalization of economic activity* (second edition). London: Paul Chapman; New York: Guilford Publications.

Donaghu, M. T. and Barff, R. 1990: Nike just did it: international subcontracting and flexibility in athletic footwear production. *Regional Studies*, 24, 537–52.

Dosi, G., Freeman, C., Nelson, R., Silverberg, G. and Soete, L. (eds) 1988: *Technical Change and Economic Theory*. London: Pinter.

Doz, Y. 1986: *Strategic Management in Multinational Companies*. Oxford: Pergamon Press.

Drucker, P. 1986: The changed world economy. *Foreign Affairs*, 64, 768–91.

Elson, D. 1988: Transnational corporations in the new international division of labour: a critique of 'cheap labour' hypotheses. Manchester Papers on Development, IV, 352–76.

Freeman, C. 1987: The challenge of new technologies. In OECD, *Interdependence and Co-operation in Tomorrow's World*, Paris: OECD, 123–56.

GATT (General Agreement on Tariffs and Trades) 1989: Services in the domestic and global economy. In *International Trade 1988–1989*, Geneva: GATT, Part III.

Gertler, M. 1988: The limits to flexibility: comments on the post-Fordist vision of production and its geography. *Transactions, Institute of British Geographers*, NS 13, 419–32.

Glickman, N. J. and Woodward, D. P. 1989: *The New Competitors: how foreign investors are changing the U. S. economy*. New York: Basic Books.

Hobsbawm, E. J. 1979: The development of the world economy. *Cambridge Journal of Economics*, 3, 305–18.

Holmes, J. 1986: The organization and locational structure of production subcontracting. In A. J. Scott and M. Storper (eds), *Production, Work and Territory: the geographical anatomy of industrial capitalism*, London: Allen and Unwin, chapter 5.

Hone, A. 1974: Multinational corporations and multinational buying groups: their impact on the growth of Asia's exports of manufactures – myths and realities. *World Development*, 2, 145–9.

Howells, J. 1990: The internationalization of R & D and the development of global research networks. *Regional Studies*, 24, 495–512.

Hymer, S. H. 1972: The multinational corporation and the law of uneven development. In J. N. Bhagwati (ed.), *Economic and World Order*, London: Macmillan, 113–40.

Johnston, R. J. 1982: *Geography and the State*. London: Macmillan.

Johnston, R. J. 1989: Extending the research agenda. *Economic Geography*, 65, 338–47.

Julius, de Anne 1990: *Global Companies and Public Policy: the growing challenge of foreign direct investment*. London: Pinter.

Kindleberger, C. P. 1969: *American Business Abroad*. New Haven: Yale University Press.

Kindleberger, C. P. 1988: The 'new' multinationalization of business. *Asean Economic Bulletin*, 5, 113–24.

Kobrin, S. J. 1987: Testing the bargaining hypothesis in the manufacturing sector in developing countries. *International Organization*, 41, 609–38.

Kostecki, M. 1987: Export-restraint arrangements and trade liberalization. *The World Economy*, 10, 425–53.

Lash, S. and Urry, J. 1987: *The End of Organized Capitalism*. Oxford: Polity Press.

Lovering, J. 1990: Fordism's unknown successor: a comment on Scott's theory of flexible accumulation and the re-emergence of regional economies. *International Journal of Urban and Regional Research*, 14, 159–74.

OECD (Organization for Economic Cooperation and Development) 1987: *International Investment and Multinational Enterprises*. Paris: OECD.

Ohmae, K. 1985: *Triad Power: the coming shape of global competition*. New York: The Free Press.

Perez, C. 1985: Microelectronics, long waves and world structural change: new perspectives for developing countries. *World Development*, 13, 441–63.

Piore, M. J. and Sabel, C. F. 1984: *The Second Industrial Divide: possibilities for prosperity*. New York: Basic Books.

Porter, M. E. (ed.) 1986: *Competition in Global Industries*. Boston, MA: Harvard Business School Press.

Porter, M. E. 1990: *The Competitive Advantage of Nations*. London: Macmillan.

Radice, H. 1989: British capitalism in a changing global economy. In A. MacEwan and W. K. Tabb (eds), *Instability and Change in the World Economy*, New York: Monthly Review Press, 64–81.

Radice, H. 1991: Capital, labour and the state. *Environment and Planning D: Society and Space*.

Sayer, A. 1986: New developments in manufacturing: the just-in-time system. *Capital and Class*, 30, 43–72.

Sayer, A. 1989: Postfordism in question. *International Journal of Urban and Regional Research*, 13, 666–95.

Schoenberger, E. 1988a: From Fordism to flexible accumulation: technology, competitive strategies and international location. *Environment and Planning D: Society and Space*, 6, 245–62.

Schoenberger, E. 1988b: Multinational corporations and the new international division of labour: a critical appraisal. *International Regional Science Review*, 11, 105–19.

Schoenberger, E. 1989: Thinking about flexibility: a response to Gertler. *Transactions, Instituite of British Geographers*, NS 14, 98–108.

Schonberger, R. J. 1982: *Japanese Manufacturing Techniques: nine hidden lessons in simplicity*. New York: The Free Press.

Scott, A. J. 1988: Flexible production systems and regional development. *International Journal of Urban and Regional Research*, 12, 171–85.

Seidler, J. 1976: *Let's Call it Fiesta*. Lausanne: Edita.

Sheard, P. 1983: Auto production systems in Japan: organizational and locational features. *Australian Geographical Studies*, 21, 49–68.

Sklair, L. 1991: *Sociology of the Global System: social change in global perspective*. Hemel Hempstead: Harvester Wheatsheaf.

Taylor, P. J. 1989: *Political Geography: world economy, nation-state and locality* (second edition). London: Longman.

Teece, D. J. 1980: Economies of scope and the scope of the enterprise. *Journal of Economic Behaviour and Organization*, 1, 223–47.

Thrift, N. J. 1989: The geography of international economic disorder. In R. J. Johnston and P. J. Taylor (eds), *A World in Crisis?: geographical perspectives* (second edition), Oxford: Basil Blackwell, chapter 2.

Thrift, N. J. 1990: The perils of the international financial system. *Environment and Planning A*, 1135–6.

Tolchin, M. and Tolchin, S. 1988: *Buying into America: how foreign money is changing the face of our nation*. New York: Times Books.

UNCTC (United Nations Commission on Transnational Corporations) *1988: Transnational Corporations in World Development: trends and prospects*. New York: United Nations.

Walker, R. 1989: A requiem for corporate geography: new directions in industrial organization, the production of place and uneven development. *Geografiska Annaler*, 71B, 43–67.

Wells, P. E. and Cooke, P. N. 1991: The geography of international strategic alliances: the cases of Cable and Wireless, Ericsson and Fujitsu. *Environment and Planning A*, 23, 87–106.

Williamson, O. E. 1975: *Markets and Hierarchies*. New York: The Free Press.

World Bank 1987: *World Development Report, 1987*. New York: Oxford University Press.

3

Social Landscapes: Continuity and Change

Susan J. Smith

In so far as it is possible to isolate a distinctively social world, the starting point for discussion must be anchored in the cultural significance of postmodernism. The term 'postmodern' has become synonymous with a new fluidity and flexibility for social life as it engages with the dynamism of the so-called 'new times'. Although interpretations of postmodernism and of its relationship with capitalism, sexism and racism vary (Harvey, 1989; Bondi, 1990; Cooke, 1990), the term is widely associated with the dissolution – at least at the level of experience – of old social categories (especially 'classes') and with the creation of new opportunities for the restructuring (or destructuring) of society.

It is often argued, and it may be true, that the postmodern world, with its ostensible classlessness and its mosaic of '3-minute' cultures, offers real possibilities for marginalized groups. It may herald new political opportunities for racialized minorities (Hall, 1988); it could promise a (for some) welcome transition from feminism to feminisms; it allows us to expose and examine in their own right the multitude of social causes once collapsed into categories like 'disadvantage' and 'deprivation' (Smith, 1989a). Certainly, one of postmodernism's attractions is its promise of 'greater democracy through its recognition of the reality of a variety of viewpoints, a plurality of cultures' (Massey, 1991:32).

Nevertheless, it can equally be argued that in order to exploit what is new about the postmodern world, it is necessary not only to document what is fluid and flexible in the accompanying social arena, but also to explain why and how certain other themes endure. Accordingly, a key concern of this chapter revolves around the possibility that while postmodernism may have added some colour to the construction of everyday life, it has not adequately challenged the basic social categories with which politicians, the public and perhaps even the majority of scholars still work. As a consequence, the thrust of my argument will not be that there is an inexorably changing world which, *ipso facto*, needs a changing discipline to study it. Rather, I will argue that precisely because there are persistencies, consistencies and continuities in the social world, because there are inequalities and injustices that we have so far failed to address adequately, and because these things are so far largely *unchanging* – despite the advent of numerous value-committed, radical, moral and applied geographies – only a radically changed discipline can hope to tackle them.

For the purposes of this chapter, therefore, the crucial leap is not from a restructured world to a changing discipline, but rather from a restructured discipline to a changing world. The difference between these is important. The first formula implicitly limits geography to an analytical role (which is not, of course, unimportant), whereas the second advocates more explicit engagement with a traditionally neglected realm of normative theory: the first approach implies detachment and impartiality in mapping the world as it is today, while the second demands a renewed commitment actively to shape the world as it *ought* to be according to some basic assumptions about human rights and social entitlements.

STRUCTURING THE SOCIAL

My starting point – that there are continuities linking the old and new times – is shared by many authors. Harvey (1989) makes a powerful and persuasive argument that there are profound continuities in the relations of production which link 'old' industrial capitalism with the project of flexible accumulation. And Massey (1991:34), albeit in a footnote, points out that 'All those lists of dualist differences between modernism and postmodernism . . . obscure the fact that an awful lot remains tediously the same'. Yet, although the broad thrust of these arguments for continuity in the sphere of political economy is widely accepted, much less attention has been paid to the possibility that continuity rather than change might also be a hallmark of the socio-cultural world. Indeed, analysts like Harvey (whose most quotable of quotes insists that postmodernism is simply the 'cultural clothing' of flexible accumulation) often imply that the very flux and variety of cultural life is required to mask the injustices of capitalism's route to a more flexible regime. A newly fluid social world – a mosaic of consumer cultures – is, effectively, cast as a prerequisite of the transition to post-Fordism. The cultural kaleidoscope of postmodernism is not, from this perspective, an illusion (as some have suggested), but rather a 'real' and necessary smoke screen for 'more of the same' in the spheres of production, investment and accumulation.

Yet ordinary people, steeped as they are in the ephemera of consumerism, seem more interested in stability than change as the twentieth century lurches to a close. Some years ago, Mary Douglas (1966) referred to that 'yearning for rigidity' which characterizes so many societies. Harvey himself admits that this has not changed. He argues that, as far as the general public is concerned, 'The greater the ephemerality, the more pressing the need to discover or manufacture some kind of eternal truth that might lie therein' (1989:292); and the greater the social fragmentation, the more potent 'the search for personal or collective identity, the search for secure moorings in a shifting world' (p. 302) – a world where localism and nationalism gain strength 'precisely because of the quest for the security that place always offers in the midst of all the shifting that flexible accumulation implies' (p. 306). In short, it seems that in the haste to reject meta-theory as a means of connecting and representing the world, social science is in danger of ignoring the extent to which people want and need to make ordered sense of their lives against a background of fragmentation and ephemera. Whether or not the world actually possesses structure and regularity, whether or not there are real ordering principles which dictate identity and shape behaviour, much of social life is acted out *as if* these principles exist. Yet our grasp on how collective ideas about a social order are developed and sustained is the first to loosen amid the excitement of analysing and conceptualizing the dynamic contours of the postmodern landscape.

In this chapter there is no space to demonstrate further what I shall take for granted – that the search for anchorage and stability continues to pervade social life. My aim is not to address the question of whether there are social continuities: there is every evidence that some fundamental divides and inequalities persist, however much they may be repositioned and renegotiated from time to time. My concern is rather to ask how it is that ordinary people (including politicians and other decision-makers) construct and experience those continuities, at a time when ideas about the class structure, consumption sector cleavages and the provisioning divide are so decidedly peripheral to the public imagination.

By way of an answer, I shall reconsider the power of metaphor – particularly of natural science metaphors – to mould perceptions and influence or justify behaviour. Metaphor is a powerful device for shaping and sharing common-sense understandings of the social world (Mills, 1982). Because of their role in establishing the 'customary vision' of a society, shaping social needs and aspirations, metaphors become vehicles for assessing what people can and, crucially, should do. Metaphors are therefore normative as well as descriptive devices, and natural science metaphors have offered a particularly powerful and enduring framework on which to construct and rationalize the emerging social order. As Bell (1990) illustrates, ideas about nature and the natural origins of the human character and behaviour are especially enduring. Historically, this natural science metaphor has been rivalled in importance in accounting for human activity only by other weighty concepts like progress and freedom.

My point here is that, despite the advent in the modern period of a wide range of powerful and persuasive social and political theories of society, despite the supposed rationality that came with modernism, and despite the advance of science itself, which has exploded so many social myths, the tenacity of natural science metaphors as an 'explanation' of, and as a source of legitimacy for, some fundamental social inequalities has been sustained, if not enhanced, by the shift to the 'new times'. I shall argue that these 'new times', which are themselves a product of technological change and innovation – a testimony to what science can achieve – encourage rather than, as we often expect, undermine our long-standing (and misplaced) tendency to justify social difference by appeal to the logic of natural science.

My argument, then, is that social and spatial boundary building is as much a feature of the postmodern world as is boundary breaking. This boundary building is not reducible to the division of labour, but neither is it a wholly welcome blossoming of cultural variety. The social bounding of the postmodern world works to legitimize, and even drive, political and economic change, but it also has a momentum of its own. And one source of this momentum is the appeal of natural science as a store of enduring 'truths' against which to measure the vagaries of human existence. My concern, then, is with the ultimate irony of a flight from progressivism: the extent to which it throws us back on those certainty-seeking models of science which, as Gould (1981:217) so often shows, are only too ready to provide an 'objectivity' for what society at large wants to hear.

SOCIAL RELATIONS, SPATIAL STRUCTURES AND THE METAPHORS OF NATURAL SCIENCE

Few dispute the extent to which, historically, the ideas of natural science have been used – with and without the sanction of natural scientists – to legitimize and reproduce inequalities and injustices in the social world. Nature, Cope argues (1985:7), has become a form of unscientific rationalization which 'projects historically specific activities, demeanours and thoughts as "natural" to all past and possible human social arrangements and relationships'. We are, nevertheless, only beginning to understand the extent to which geography itself is implicated in that process. Perhaps the best account is given by Livingtone (1991), who shows how geographers drew scientific climatology into the process of race categorization, domination and exploitation which underpinned the colonial project of nineteenth-century Europe. He shows that 'the idea that climate had stamped its indelible mark on racial constitution not just physiologically, but psychologically and morally, was a motif that was both deep and lasting in English-speaking geography' (p. 10). This motif informed a naturalization of human values which allowed human potential, morals and social worth to be gauged in climatic terms. This 'language of climate's moral imperatives' was used as a legitimizing discourse for the colonists as they sought to erect 'crucial boundaries between civilisation and barbarism, between the white and black races, and of course, between virtue and vice' (p. 24).

The problem for social science today, however, is that the use of natural science metaphors as legitimizing discourses is not restricted to the past. It is part of a continuing process of renegotiation required as the social world is jostled by the exigencies of politics and economy. Some consequences are itemized by Cope (1985:7):

> in response to a women's movement wanting the creation of new gender relations, present family structures are characterised as natural and to be found across all history. In response to struggles to change the unequal relation of human groups of different geographic origin and with different physical features, there arise theories about the natural superiority of 'race' . . . In response to bitterness about the unequal results of schooling for industrialism emerge arguments that intelligence . . . is biologically inheritable, a gift or penalty of nature. In response to calls for the end of the arms race, it is argued that aggression is natural. Even our co-operative activity can be put down to the functioning of a selfish 'gene'.

Postmodern life remains shot through with such discourses, and they continue to oppress. Indeed, it can be argued that they have gained new life through the industrial, political and welfare restructurings that comprise the 'new times'. The examples of the sustained naturalization of gender and 'race' differences well illustrate this point.

Women and Work

Ideas about women's inherent suitability for certain kinds of work have always informed, and been informed by, the naturalization of gender differences (the process by which socially constructed differences between men and women are accounted for in biological terms, thus emphasizing their inevitability rather than their openness to change). These differences are enshrined in the social contract of most countries of the developed world (Pateman, 1988). Nevertheless, times are changing: equal pay legislation has been enacted to challenge the practices that devalue women's work, and anti-discrimination legislation demands a more open set of employment policies. These have not succeeded in eradicating the inequalities that flow from occupational segregation and gender discrimination (Fincher, 1989), but symbolically, and in practice, their existence might be expected to undermine the natural science metaphor which has traditionally so readily legitimized income differentials.

Today, the work that women do is recognized as playing an important part in the economic restructurings of the 'new times' on a global as well a national scale. Yet in examining this restructuring, we find that ideas about women's supposedly 'natural' suitability for work in different parts of the production process have been reworked rather than discarded. These assumptions continue effectively to legitimize women's unequal opportunities, pay and conditions, even though the economic context has changed (Wekerle and Rutherford, 1989).

Ironically, at a time when female labour is widely required to bolster the casualized periphery of the post-Fordist economy, it is becoming hard to overstate the extent to which supposedly natural differences between men and women in their ability or suitability to perform certain tasks is still drawn on by employers to 'explain' the gender division of labour in the paid workforce. Elson and Pearson (1981), for instance, show how themes related to 'nimble fingers', 'natural manual dexterity' and 'patience' are constructed as female skills, and are inserted into a 'natural' hierarchy beneath male skills in order to sustain, and account for, women's secondary status in the third world labour force (a labour pool which now drives the production process). In the same vein, Metter

(1986) discusses the feminization of the British electronics industry – an industry in which women are valued for their 'natural manual dexterity' and 'patience'. Metter shows how these 'feminine' skills are constructed in an ideologically biased way to legitimize ideas about a 'natural' hierarchical distinction between the value (and therefore pay and conditions) of men's and women's work.

It is a short step from this type of analysis to argue that, in a country like Britain, ideas about the natural location of women's work – the home – have legitimized, if not facilitated, a shift in women's work patterns from the regulated to the unregulated sectors of the labour market, or from the open to the hidden economy (McDowell and Massey, 1984). This same discourse has often helped drive women out of the workforce altogether in a political climate which favours a return to 'Victorian values' in the wake of rising unemployment. Bleir (1984) takes up this point by tracing the flexible yet persistent role of scientific metaphor in developing a myth of female inferiority, which has been consistently drawn upon to 'explain' and legitimate women's subordinate economic position in the western world.

In short, it seems that with the advent of flexible accumulation, and the restructuring of the labour market that goes with this, the naturalization of gender differences, which secures at least some social legitimacy for a particular division of labour, is not a relic of the modern world. Rather, it plays a crucial part in both the geography and sociology of those economic restructurings which we associate with a shift to the 'new times'.

'Race' and Politics

A similar set of examples can be drawn on to illustrate the sustained salience of 'race' in a period where a traditional appeal to the ordering principles dictated by the physical and biological sciences has no demonstrable foundation. Racialization is generally recognized as a process whereby somatic traits are overlaid with presumptions of natural origin, and infused with social significance. It refers to the social construction of human races, and to the naturalization of these categorizations. I have argued elsewhere that, in Britain in the last fifty years, the process of racialization – the assumption that human races are real – has informed the construction of selective immigration laws and has legitimized the unequal division of residential space between 'black' and 'white' Britons (Smith, 1989b). I have shown, too, how these themes of exclusion and segregation map on to a geography of racism, and I have argued that this geography helped mask the inequalities that flowed from, and indeed fuelled, Britain's post-war economic reconstruction. The naturalization of racial differences certainly has a role alongside the naturalization of gender difference in the regulation of industrial capitalism.

But there is more to it than this. For changing ideas about 'race' (and gender too) have also been at the heart of the political reorderings which have been as prominent in the transition to the 'new times' as has economic upheaval. As Barker (1981) ably shows, the shift towards neo-conservatism which has characterized so much of the western world in the 1980s has been accompanied by the tendency for 'respectable' politicians to retreat from the crude language and imagery of 'race'. They employ instead the postmodern language of pluralism, diversity and apparent tolerance of difference. This is a world where the dichotomy 'black/white' gives way to a deliberate cultural and ethnic pluralism. But it is also a backcloth for what might be called the racialization of culture: 'the conflation of "race" with culture, leading to the categorization and subjugation of individuals on the basis of how they are presumed to act, what they are presumed to think, and where their religious, linguistic and national loyalties are presumed to lie' (Smith, 1992:137).

The presumption that lies at the heart of this line of thinking is that cultural difference is nature's product. Underlying the new veneer of cosmopolitanism there is, accordingly, a strong adherence to ideas about the essentially natural origins of cultural differences. It is significant, therefore, that these 'natural' cultural boundaries are not drawn round just any set of shared meanings. Rather, new debates on ethnicity and culture are shot through with euphemistic references to old 'race' categories. This is recognized by Cope (1985:18), who points out that 'naturalising culture, culturalising structure, and thus making it seem that the world is the way it is in part because of primordial immutable traditions or ethnicities, is a polite new form of racism'.

Once defined, these 'natural' differences, whether they are described in terms of 'race', 'ethnicity' or 'culture', ensure that judgements of superiority and inferiority are expressed through the reasonable concept of difference, so that a friendly face conceals the uncomfortable facts of inequality. But today, no less than in the past, these ideas of difference are drawn on to define and strengthen distinctive forms of nationalism in the face of a reordering of world politics (Gilroy, 1987; Miles, 1987). The case of the UK is just one example of how, by packaging social categorizations based on skin colour or imputed national origins as a celebration of multiculturalism, politicians, the public and many analysts deflect attention from the racisms which bolster the force of nationalism.

Today's implicit appeal to natural science as the source of social differentiation cloaks assertions of power in benign expressions of identity, but it legitimizes the geography of inequality every bit as effectively as older, more explicit references to a natural racial hierarchy.

THE MEDICALIZATION OF SOCIAL LIFE

Having briefly used the examples of an entrenched naturalization of gender and 'race' differences to illustrate the sustained salience of a natural science model of social order throughout the processes of economic transition and political reordering, I shall now dwell in a little more detail on a third natural science metaphor. My aim is to illustrate how the medicalization of social life informs and is informed by a restructuring of welfare – a third integral element of the postmodern 'new times'.

Medicalization refers to the drawing of boundaries around social groups on the basis of presumed health, illness or susceptibility to disease. Such boundaries are overlaid with attitudinal, behavioural and territorial markers, and may be used as criteria in determining the differential apportionment of goods and services. Crucially, medicalization is about a process of social categorization for the purposes of *rationing* resources and *controlling* personal and public space. The history of this incorporation of medical metaphors into the project of domination and control is increasingly well documented. An explicit appeal to medical knowledge has, in particular, been used to control the life-spaces of women and racialized minorities throughout the modern period.

The essays collected by Macleod and Lewis (1988), for instance, show how medicine was used as an agency of cultural domination in the colonial period: it was offered as proof of imperial superiority and used as a vehicle for the exercise of social control over colonized peoples. The spectre of uncontrolled infectious disease among 'native' populations was also used by colonizers to justify the segregation and subjugation of colonial populations. This is well illustrated in Frenkel and Western's (1988) account of the development of segregationism in Sierra Leone. An entrenched image of 'racial' outsiders as the harbingers of not only physical but also moral contamination was also influential in shaping the policies and institutions of segregation in South Africa (Swanson, 1977).

In a similar vein, Proctor's (1988) powerful account of 'racial hygiene' in Nazi Germany shows how potent a role medical science played in shaping a racial ideology. Proctor shows how medicine interacted with the project of national socialism to exemplify the broader generalization that 'if people can be convinced that the social order is a natural order, and that the misery (or abundance) they find around them derives from the will of God or Nature or both, then attention can be diverted from those parts of the social order that are the true source of that misery (or abundance)' (p. 2). The historical interlinking of medicine and racism is further explored by Littlewood and Lipsedge (1989), who show how psychiatry, by medicalizing the mental health problems of racialized minorities, has helped translate the outsider status associated with ideas about 'race' into the marginal status of people labelled 'mentally ill'.

The moral environmentalism so instrumental to early ideas of racial difference and spatial segregation in the public sphere has also been central to the control of women's life-space, especially in the private sphere. Mort (1987) attributes some of our most enduring constructions of gender difference to the social hygiene movement of the early twentieth century. This movement rolled together ideas about health, morality, sexual behaviour and passion, and produced new representations of sex difference organized around the themes of (male) normality and (female) abnormality.

The progressive medicalization of women's lives is taken up elsewhere. Ehrenreich and English (1979, 1981) show how early myths of female frailty have limited women's social opportunities, and they go on to illustrate that the medicalization of this syndrome not only debarred women from practising as doctors, but also qualified them as patients and so boosted the coffers of a newly developing, male-dominated medical profession in the late nineteenth century. Cayleff (1988) traces this gender bias in the medical conceptualization, diagnosis and treatment (and associated common-sense understandings) of nervous disorders across the whole time–space spectrum from medieval Europe to twentieth-century North America. She points to the remarkable consistency with which a 'view of women's predisposed susceptibility to nervous debility transcended medical and scientific knowledge to include assertions that stemmed from deeply held beliefs about female and male "natures" and the acceptable parameters of women's behaviour and influence' (p. 1205). Exploring a related area, Clarke (1983) shows how the medicalization of childbirth which gave rise to its own branch of medicine – obstetrics and gynaecology, which is still dominated by men – has been harnessed to the social control of women. This theme is taken up by Abel and Kearns (1990), who show how the appropriation of maternity services into hospital settings has extended control into the most intimate spaces of women's lives and deprived them of autonomy and security in some key areas of family life.

The power of medicalization as an ordering principle for social life is not, then, in dispute. Neither, in any serious way, is its tenacity. Cayleff (1988) concludes her review with the sobering observation that 'gender ideologies still largely inform the illness labeling, medical diagnosis and management of woman's physiology' (p. 1205). Spallone (1989) argues in the same vein that the new reproductive technologies compromise women's integrity, rights and freedom every bit as much as the earlier population promotion and regulation policies. My own view is that medicalization may be one of the most powerful and persuasive natural science metaphors contributing to the ordering and reordering of society. From the perspective of this chapter, it is important because it is implicated not only in defining and controlling gender and 'race' difference, but also in positioning and discriminating among the many other social groupings – the elderly', the 'physically disabled' and so on – emerging through the process of postmodernization.

The medical metaphor currently informs many areas of policy and practice. For instance, it controls access to certain forms of employment: and as science itself advances, the potency of presumed health status as a barrier to access to the core sectors of the post-Fordist labour market seems set to increase. In the USA, for example, where black Americans with sickle-cell trait used to be debarred from the Air Force, tests for AAT-deficiency (a genetically controlled shortage of an enzyme that detoxifies tar, and which may arguably predispose its incumbents to emphysema) are now commercially available to employers. Likewise, the medical metaphor is increasing rather than decreasing in importance as a means of regulating political frontiers and other territorial boundaries. Gordon (1983) shows how, during the late nineteenth and early twentieth centuries, health checks were used as immigration controls by the developed countries, ostensibly to stop the import of infectious disease but in reality to limit non-white immigration. Taking the example of the UK, Gordon makes the further claim that such controls are still used: not to protect public health but to discriminate among those British citizens who are, and those who are not, eligible to live with their families in the UK. An appeal to medical knowledge is made to determine, for instance, whether immigrant 'fiancées' are virgins, whether immigrant children are the 'natural' offspring of sponsoring parents, and whether such children are young enough to qualify as dependants. Gordon's point has more general relevance: 'medical controls have a special significance in that they appear to be based on the scientific judgement of objective facts, and therefore not tainted by more political considerations' (p. 17).

It is in this light that we can begin to understand the relevance and potency of the medical metaphor in the process of rationing resources and regulating deviance in a restructured welfare state. The potential extent and consequences of the medicalization of public policy is documented by Binney, Estes and Ingman (1990) in their study of community-based services for older people in the USA. This research shows how community-based service provision has been steadily restructured to favour medical intervention over social support. This has occurred to such an extent that the possibilities for providing comprehensive care are often frustrated, and socially oriented services – for many, a prerequisite for independent living – are frequently under threat. This example indicates that medicalization is not about implementing the welfare ideal of providing for individuals according to their own distinctive needs (medical or otherwise). Rather, it can be argued that the consequences of medicalization may fundamentally compromise (or may be used to legitimize policies which fundamentally compromise) this welfare ideal, as cutbacks in public expenditure are traded for tax concessions in a bid to roll back the state and free up the market-place. In documenting this, I am concerned not with the crude medical metaphors of the colonial period but with a more subtle intrusion of quasi-medical categories into the process of welfare restructuring in the late twentieth century. In particular, I consider how medicalization impinges on social policy in a way that helps define and separate the 'deserving' and 'undeserving' poor in order to accommodate the residual model of welfare now embedded in our thinking about welfare transfers.

Housing and Health: Medicine and the Urban Order

To illustrate the point, I take the example of British housing policy, both because housing has been at the leading edge of welfare restructuring in this country (Clapham, Kemp and Smith, 1990) and because the reorganization of living space this has entailed is of particular relevance to the project of geography and can be directly linked to some of the processes of medicalization with which I am concerned (Smith, 1990a).

The British housing system was, for many years, a cornerstone of the country's welfare state. It epitomized the widely accepted principle that a civilized society collectively provides for those disadvantaged in a market system by factors, like illness, which are beyond individuals' control. Thus from as early as the 1930s, but especially from the 1950s, public housing has acted as a safety net for people whose incomes are depressed by ill-health. Systems of housing allocation therefore include mechanisms for giving people with medical needs priority in access to, and transfer within, the subsidized rented stock. Housing provision is therefore theoretically health selective in favour of sick people: people with health problems traditionally have the pick of public housing space. And, if anything, the legislative impetus for this health selectivity has been enhanced in recent years (Smith, 1989c, 1990b), so that the principle of awarding priority access to mainstream housing on medical grounds is now firmly embedded in the housing management practices of most local authorities.

However, the public sector is changing. Britain, like North America (Wolch, 1989) and Australia (Fincher, 1989), has undergone a period of welfare restructuring. In the UK, this shift from the state to the market has been spearheaded by the reorganization of housing provision. Until very recently, therefore, housing has borne the brunt of policies aiming to reduce public expenditure in return for cuts in taxation. As Forrest and Murie (1988) show, it is in the sphere of housing provision and management that Britain has most effectively relinquished a model of provisioning based on direct state intervention (through the provision of subsidized rented dwellings) to a model based on market principles (through the encouragement of owner-occupation). Irrespective of whether this is a good or a bad thing, it has consequences which are relevant to the theme of this chapter.

The transition from state to market in housing provision has been achieved primarily through the sale of public rented homes to sitting tenants, at discounted prices and with advantageous mortgage arrangements. New capital investment has not kept pace with sales: indeed, councils have not usually been allowed to use receipts from sales to build replacement stock. Because the pattern of purchase has been uneven (suburban houses have sold more readily than inner-city flats; dry homes in good repair have proved a more attractive investment than damp homes in poor condition), the council rented stock which remains after a decade of the 'right to buy' has decreased in size, diminished in quality and become more restricted in its geography.

Observers link this residualization of council housing with an associated marginalization of council tenants (Forrest and Murie, 1987), since those tenants most able to buy are the better-off, middle-aged, securely employed families. Those who remain as renters tend to be very young or old, and, for the most part, benefit-dependent (for a variety of reasons, including unemployment and poor health). This marginalization is evident in the increasing sociotenurial polarization of the housing system as a whole (Bentham, 1986; Robinson, 1986). Council housing has become more explicitly the welfare arm of the housing system, so attracting more applications from people with medical needs, even as medical priority is forced to compete with a growing range of other priority claims (Smith, 1990b; Connelly and Roderick, 1992). The system is therefore under pressure and the ability of the mainstream housing stock to accommodate people with a range of general health problems is increasingly questionable (Parsons, 1987). Even those who do secure access to this part of the housing system have no guarantee of a healthy home (though certain adapted dwellings are exempt from the sales policy). It is at this point that our attention must shift from the welfare ideal (of matching housing accommodation to needs) to the process of medicalization (using ideas of health and disease to determine who does and does not deserve public subsidy).

The restructuring of housing provision is implicated in the medicalization of residential space in at least three ways. First, it encourages the incorporation of medical criteria into housing needs assessments, bringing clinical judgements to bear on the managerial problem of matching housing applicants to the available stock of dwellings. Second, it has underpinned the development of 'special' housing – separate living space – for people with certain health problems. Finally, it has a bearing on the health profile of homeless people, who, by virtue of their exclusion from mainstream housing services, occupy a quite different niche in the medicalized landscape.

1 Health problems and housing management Many local authorities delegate the assessment of all applications for housing priority on medical grounds to a health professional (usually a public health physician, but sometimes an occupational therapist or local general practitioner). In a recent survey of one in three English local authorities, 68 per cent of housing departments claimed always to consult a medical advisor (and a further 16 per cent sometimes consult). In half the authorities concerned, the advice of the medical advisor is regarded as binding, and in a further 40 per cent of authorities the advice is 'usually followed' (Smith, McGuckin and Walker, 1991).

This appeal to medical expertise is not intrinsically problematic. It may, for instance, provide a means of ensuring confidentiality to applicants with health problems. But it is a hallmark of institutional medicalization – a process usually deemed to occur 'when the physician is elevated to the position of "gatekeeper" to authorize eligibility for services' (Binney, Estes and Ingman, 1990:762). In the field of housing provision, this kind of procedure has been criticized for translating housing management issues into questions of clinical judgement – a judgement which may be unnecessary or inappropriate given that the majority of housing outcomes are stock-led rather than needs-related (Parsons, 1987). The problems are only compounded when the health professionals concerned have relatively little knowledge of the housing system they are dealing with (Kohli, 1986).

Whatever the merits or otherwise of incorporating clinical judgements into housing needs assessments, the consequences of the medicalization of housing management decisions may be far-reaching. Although this system allows housing priority to be determined on the basis of health problems, it is not clear to what extent the health problems most likely to secure housing priority are actually housing-related, and it is not known to what extent rehousing is an appropriate health intervention in the majority of cases. Moreover, at a time when people with medical priority are competing with a range of other legitimate priority claims on the public housing system, there may be little justification for separating health from other welfare needs in the allocation of a limited quantity of living space (and some local authorities have recognized this by replacing their medical needs categories with a broader special needs list). On the other hand, as councils' better-quality properties need to be more stringently rationed in the face of increasing demand and decreasing supply, it is possible that definitions of health and illness – and their association with different levels of housing priority – will be increasingly important as measures of more or less eligibility for public housing services (and for the better-quality parts of public sector housing space). Those least eligible may slip through the welfare net altogether (see section 3 below); those most eligible are likely to be directed toward the 'special' housing spaces considered in the next section.

2 'Special' Housing Space The difficulties the public sector now has in providing for general medical needs through mainstream housing allocations has produced an alternative model of housing provision. This has resulted in the development of a form of 'medicalized space' geared to the needs of the unambiguously 'deserving' poor. The production of 'special' medical space is epitomized in the advent of the special housing movement, which demands the standardization of illness for management purposes in ways which Turner (1987) finds at the heart of the medicalization of society.

'Special' housing is designed for people who need special home adaptations or special forms of care if they are to live independent lives in the community. It is ideal for people with certain medical or social support needs, and it has developed at least partly in response to policies of deinstitutionalization. Special housing includes: sheltered/amenity housing for older people; supported accommodation for people with learning difficulties or mental illness; and wheelchair or mobility standard housing for people with walking, stretching and reaching difficulties.

This model of provision has secured excellent homes for relatively small groups of 'elderly', 'mentally ill', 'mentally handicapped' and 'disabled' people. In practice, however, the current enthusiasm for special housing marks the end of the use of the public sector to meet general housing needs. It is a product of welfare restructuring and it is concerned not merely with targeting resources but also, and crucially, with rationing them (Clapham and Smith, 1990). Special housing is significant, then, not because of the amount of shelter supplied (which is relatively small) but because it is one of the few areas of public housing still receiving capital investment. In fact, we know that many people, even with supposedly legitimate special needs, never get homes in special schemes, and by comparing the geography and sociology of mainstream and special provision the social costs of the residual model of welfare from which the 'special' model flows can readily be appreciated.

First, the 'special' model tends to segregate client groups into 'special' space. Special housing is often segregated and has been criticized for providing mini-institutions in community settings. By packaging certain buildings with certain forms of care, this model of provision can also limit the locational options of those who only require the care or dwelling element of the package. Second, the cost of gaining access to special housing may be the sacrifice of individual identity to collective stigma. By catering in a standardized way to grouped needs, special accommodation may reinforce labels like 'elderly', 'mentally handicapped' and so on, even though many older people are fit and well, and many learning difficulties are slight. Finally, it is necessary to acknowledge that housing provision has often been bound up with two opposing models of health care; that concerned with control and containment and that oriented towards disease prevention and health promotion (Smith, 1990a). There has always been a tension between these models, and there is a danger that special housing, which began in the spirit of prevention and promotion, is – as it becomes infused with medical metaphor – sliding towards the control/containment end of the spectrum.

The dilemma this raises is epitomized in the tensions running through the AIDS and housing movement. Because local authorities have few policies for admitting people with HIV/AIDS to mainstream housing systems, such people find they have to bargain for extra – new – resources to secure adequate and properly serviced accommodation. The current model for levering such resources from the closely guarded public purse is the 'special' one – but this runs the risk of isolating/stigmatizing an already persecuted group (at the same time as it deflects attention away from the inadequacies of provision via the mainstream housing services). The dangers of this model are heightened because the theme of quarantine, which Altman (1986:18) identifies as 'probably the oldest public health measure', is ever-present in the current hysteria over the management of AIDS and HIV; and the social reactions evoked by this disease are strikingly similar to those that motivated past quarantine efforts (Musto, 1988).

This theme of the control and containment of people with HIV is evident on a variety of spatial scales. On the macro-scale there are immigration restrictions to prevent the entry of people who are HIV-positive to the USA, and in Australia calls have been made to confine gay men to particular islands (Altman, 1986). The same sentiments have appeared in regional policy: in 1983 the then-chief of infectious diseases in California is alleged to have invoked the idea of segregation as a means of dealing with 'recalcitrant' AIDS patients; in Britain, legislation passed in 1985 allows local authorities to keep people with AIDS in hospital if they are deemed a risk to others. The dangers of this line of thinking when extended to the micro-scale – into policies concerned with the apportionment of living space – can be seen in Sweden, where certain HIV-positive people (those classed as promiscuous) live, are cared for and are contained within special dwellings. What we see epitomized in the experiences of people with HIV/AIDS is the more general logic of the special housing model which, for all its positive achievements, leans towards a medical division of residential space, and lends weight to a growing social divide between those who are deemed fit, healthy and productive, and those known as sick, frail or benefit-dependent.

3 Homelessness and health Homelessness is increasing rapidly in Britain, and this – like the advent of special housing – can be linked directly to the reorganization of housing provision (Murie, 1988). Homeless people have a distinctive health profile, which is readily related to their harsh and hazardous living space and to their demonstrably limited access to primary medical care (Smith, 1989c). This profile is increasingly prompting homelessness to be thought of in medical terms, just as it was once thought of as an index of criminality or a consequence of individual pathology (Shanks and Smith, 1991). Homelessness, which was in the past criminalized, is increasingly medicalized. Homeless people can thus be said to occupy a second kind of medicalized space – that available to the so-called 'undeserving poor', who have not been able to exercise their statutory right to shelter. (Usually, in the British example, homeless people who do not qualify as unintentionally homeless, in priority need and having a local connection, are excluded from the public sector: Clapham, Kemp and Smith, 1990.)

The image of poor health as a corollary of homelessness is, from the perspective of public policy, deflecting attention away from the causes of homelessness (in the housing system) and towards its effects (here, its consequences for public health). It is, of course, important that inequalities in access to primary medical care between housed and homeless populations are addressed, and to this end an emphasis on the role of the health services in tackling the health of homeless people is welcome. On the other hand, there is a very real possibility that many people with health problems are routinely excluded from public sector housing (because of the way the bureaucratic rules defining eligibility admit some kinds of health problem and exclude others in order to ration a diminishing pool of suitable accommodation) at a time when there is an affordability crisis in the private sector. This in itself has a bearing on the problematic health profile of homeless people today. The irony, then, is that if the rules by which the housing system operates are interrogated, there is every indication that a growing segment of the health profile of homeless people is accounted for by the process of health selection out of housing and into the street (Smith, 1989c; Shanks and Smith, 1991). In short, the restructured housing system has an active role in shaping the health profile of homeless people, even as public responses to that profile – by constituting the problem in terms of health care rather than housing services – are effectively protecting the housing system from fundamental reform by diverting attention towards the gaps in health service provision.

The 'new times' are signalled by a process of welfare restructuring as well as by economic transition and political reordering. In Britain, the shift from state to market provisioning has changed the face of public sector housing space, and the process has been both informed and legitimized by the medical metaphor. By drawing clinical boundaries around social groupings, it has been possible to reconstitute the distinction between a deserving and an undeserving poor, and shed the light of scientific certainty – of natural inevitability – around a new urban order which is, nevertheless, shot through by old inequalities.

CONCLUSION

The world I have looked at, and it has been a partial view, is undoubtedly changing. But my theme has been that, despite the cultural fluidity of these postmodern times, the processes of social categorization go on, and the basic axes of social inequality remain (notwithstanding some rescaling of their dimensions). The question at the heart of my discussion is therefore this: how and why do we, reasonable people, tolerate enduring inequality?

We do so, I have suggested, because the sources of the inequalities, and the social boundaries concerned, are acceptable at the level of common sense. One of several metaphors that make them seem plausible is the continued naturalization of social life. An appeal to that most rational of certainty-seeking human endeavours, natural science, wins legitimacy for the reproduction of inequality, and for the control and oppression of marginal groups. The point here is that for all our sophisticated social and cultural theory, for all our understanding of economic change and the processes of legitimation, the appeal of a scientific logic to social differences has not gone away. On the contrary, the very technical advances that gave the 'new times' their economic impetus also inform the naturalization of gender differences in a restructured workforce, the racialization of culture in a reorganized political world, and the medicalization of social needs in a restructured welfare state.

What can be done? That is the topic of another essay – and of other chapters in this book. I conclude, nevertheless, with the observation that in order to engage in a changing world, we need a well-developed sense not only of what that world looks like now, but also of what it *should* look like for future generations. Without a better sense of the value of normative theory, without more willingness to enter the politics of prescription, geography is powerless to challenge the subtle ideologies that legitimize enduring social inequalities. It is not enough simply to know how and why the world is changing; if geography has any relevance in the postmodern world, it must have something new to say about how and why the world *should* change.

REFERENCES

Abel, S. and Kearns, R. 1990: Birth places: a geographical perspective on planned home birth in New Zealand. Unpublished paper, University of Auckland, Departments of Anthropology and Geography.

Altman, D. 1986: *AIDS and the New Puritanism*. London: Pluto.

Barker, M. 1981: *The New Racism*. London: Junction Books.

Bell, M. 1990: Class, community and nature in an English village. Draft of a Ph.D. dissertation, Yale University.

Bentham, G. 1986: Socio-tenurial polarization in the United Kingdom 1953–83: the income evidence. *Urban Studies*, 23, 157–62.

Binney, E. A., Estes, C. L. and Ingman, S. R. 1990: Medicalization, public policy and the elderly: social services in jeopardy?. *Social Science and Medicine*, 30, 761–71.

Bleir, R. 1984: *Science and Gender*. Oxford: Pergamon Press.

Bondi, L. 1990: Feminism, postmodernism and geography: space for women?. *Antipode*, 22, 156–67.

Cayleff, S. E. 1988: Prisoners of their own feebleness?: women, nerves and western medicine – a historical overview. *Social Science and Medicine*, 26, 1199–208.

Clapham, D. and Smith, S. J. 1990: Housing policy and 'special needs'. *Policy and Politics*, 18, 193–205.

Clapham, D., Kemp, P. and Smith, S. J. 1990: *Housing and Social Policy*. Basingstoke: Macmillan.

Clarke, J. 1983: Sexism, feminism and medicalism: a decade review of literature on gender and illness. *Sociology of Health and Illness*, 5, 62–82.

Connelly, J. and Roderick, P. 1992: Medical priority for rehousing: an audit. In S. J. Smith, A. McGuckin and R. Knill-Jones, (eds), *Housing for Health*, London: Longman, 73–91.

Cooke, P. 1990: Modern urban theory in question. *Transactions, Institute of British Geographers*, NS 15, 331–43.

Cope, B. 1985: Racism and naturalness. *Social Literacy Monograph 14*, Centre for Multicultural Studies, University of Wollongong.

Douglas, M. 1966: *Purity and Danger: an analysis of the concepts of pollution and taboo*. London: Routledge & Kegan Paul.

Elson, D. and Pearson, R. 1981: 'Nimble fingers make cheap workers': an analysis of women's employment in third world export manufacturing. *Feminist Review*, 7, 87–107.

Ehrenreich, B. and English, D. 1979: *For Her Own Good: 150 years of the experts' advice to women*. New York: Anchor Press/Doubleday.

Ehrenreich, B. and English, D. 1981: The sexual politics of sickness. In P. Conrad and R. Kern (eds), *The Sociology of Health and Illness*, New York: St Martins.

Fincher, R. 1989: Class and gender relations in the local labor market and the local state. In J. Wolch and M. Dear (eds), *The Power of Geography*, Boston, MA: Unwin Hyman, 91–117.

Forrest, R. and Murie, A. 1987: The pauperization of council housing. *Roof*, Jan.–Feb., 20–3.

Forrest, R. and Murie, A. 1988: *Selling the Welfare State: the privatisation of public housing*. London: Routledge & Kegan Paul.

Frenkel, S. and Western, J. 1988: Pretext or prophylaxis? the malarial mosquito and racial segregation in a British tropical colony. *Annals of the Association of American Geographers*, 78, 211–28.

Gilman, S. L. 1988: *Disease and Representation*. Ithaca, NY, and London: Cornell University Press.

Gilroy, P. 1987: *There Ain't No Black in the Union Jack*. London: Heinemann.

Gordon, P. 1983: Medicine, racism and immigration control. *Critical Social Policy*, 3, 6–20.

Gould, S. J. 1981: *The Mismeasure of Man*. New York: Norton.

Hall, S. 1988: Brave new world?. *Marxism Today*, 32, 24–9.

Harvey, D. 1989: *The Condition of Postmodernity*. Oxford: Basil Blackwell.

Kohli, H. 1986: Medical housing 'lines'. *British Medical Journal*, 293, 370–2.

Littlewood, R. and Lipsedge, M. 1989: *Aliens and Alienists* (second edition). London: Unwin Hyman.

Livingstone, D. N. 1991: Climate's moral economy: science, race and place in post-Darwinian British and American geography. Paper presented to the conference on *Geography and Empire: critical studies in the history of Geography*, Queen's University, Kingston, Ontario.

MacLeod, R. and Lewis, M. (eds) 1988: *Disease, Medicine and Empire: perspectives on Western medicine and the experience of European expansion*. London: Routledge.

Massey, D. 1991: Flexible sexism. *Environment and Planning D: Society and Space*, 9, 31–58.

McDowell, L. and Massey, D. 1984: A woman's place?. In D. Massey and J. Allen (eds), *Geography Matters!*, Cambridge: Cambridge University Press, 128–47.

Metter, S. 1986: *Common Fate, Common Bond*. London: Pluto.

Miles, R. 1987: Recent Marxist theories of nationalism and the issue of racism. *British Journal of Sociology*, 38, 24–41.

Mills, W. J. 1982: Metaphorical vision: changes in western attitudes to the environment. *Annals of the Association of American Geographers*, 72, 237–53.

Mort, F. 1987: *Dangerous Sexualities: medico-moral politics since 1830*. London: Routledge & Kegan Paul.

Murie, A. 1988: The new homeless in Britain. In G. Bramley, K. Doogan, P. Leather, A. Murie and E. Watson (eds), *Homeless and the London Housing Market*, Occasional Paper 32, Brown: School for Advanced Urban Studies.

Musto, D. F. 1988: Quarantine and the problem of AIDS. In E. Fee and Fox (eds), *AIDS: the burdens of history*, Berkeley, CA: University of California Press, 67–85.

Parsons, L. 1987: Medical priority for rehousing. *Public Health*, 101, 435–41.

Pateman, C. 1988: *The Sexual Contract*. Cambridge: Polity Press.

Proctor, R. N. 1988: *Racial Hygiene: medicine under the Nazis*. Cambridge, MA: Harvard University Press.

Robinson, R. 1986: Restructuring the welfare state: an analysis of public expenditure, 1979/80–1984/85. *Journal of Social Policy*, I5s, 1–21.

Shanks, N. and Smith, S. J. 1991: Public policy and the health of homeless people. *Policy and Politics*, 20, 35–46.

Smith, S. J. 1989a Social geography: social policy and the restructuring of welfare. *Progress in Human Geography*, 13, 118–28.

Smith, S. J. 1989b: *The Politics of 'Race' and Residence*. Cambridge: Polity Press.

Smith, S. J. 1989c: Housing and health: a review and research agenda. Discussion paper 27. Glasgow: Centre for Housing Research.

Smith, S. J. 1990a: AIDS, housing and health. *British Medical Journal*, 300, 243–4.

Smith, S. J. 1990b: Health status and the housing system. *Social Science and Medicine*, 31, 753–62.

Smith, S. J. 1992: Residential segregation and the politics of racialisation. In M. Cross and M. Keith (eds), *Racism and the Postmodern City*, London: Unwin Hyman.

Smith, S. J. McGuckin, A. and Knill-Jones, R. (eds) 1992: *Housing for Health*. London: Longman.

Smith, S. J., McGuckin, A. and Walker, C. 1991: Housing provision for people with medical needs. Paper for the conference on *Unhealthy Housing: the public health response*, December.

Spallone, P. 1989: *Beyond Conception: the new politics of reproduction*. Basingstoke: Macmillan Education.

Swanson, M. W. 1977: The sanitation syndrome: bubonic plague and urban native policy in the Cape Colony, 1990–1901. *Journal of African History*, 18, 387–410.

Taylor, S. M. 1989: Community exclusion of the mentally ill. In J. Wolch and M. Dear (eds), *The Power of Geography*, Boston, MA: Unwin Hyman, 316–30.

Turner, B. S. 1987: *Medical Power and Social Knowledge*. London: Sage Publications.

Wekerle, G. R. and Rutherford, B. 1989: The mobility of capital and the immobility of female labour: responses to economic restructuring. In J. Wolch and M. Dear (eds), *The Power of Geography*, Boston, MA: Unwin Hyman, 139–72.

Wolch, J. R. 1989: The shadow state: transformations in the voluntary sector. In J. Wolch and M. Dear (eds), *The Power of Geography*, Boston, MA: Unwin Hyman, 197–221.

4

Ends, Geopolitics and Transitions

Graham Smith

With the demise of state socialism and the end of the Cold War, a common consensus has emerged in both popular and scholarly discourses that we have come to the end of an historical epoch. For the New Right, the end of the Cold War represents nothing less than the 'end of history', and the victory, reflected in the revolutionary years of 1989 and 1991 in Eastern Europe and the Soviet Union, for economic and political liberalism, a triumph for the west that presages the global universality of capitalism and the end of Communism. It is seen as representing the ascendancy of the 'universal homogeneous state', defined as liberal democracy in the political sphere combined with access to the material and cultural way of life associated with late capitalism (Fukuyama, 1989).

It is an interpretation not that far removed from the standard liberal viewpoint. Here modernization has been the victor, a synonym for the end of totalitarianism in whatever state form, for the emergence of constitutional democracy, and for economic and social pluralism. The 'end of ideology', as Bell (1961, 1973) rather hastily put it, has finally come true. The 1989 revolutions and the 'second Russian revolution' demonstrated that 'actually existing socialism' failed either to realize the importance of individual freedoms and community self-government as reflected in Marx's early humanistic writings, or to fulfil, through central state planning, material conditions unmatched by market economies. In short, the 1989 and 1991 revolutions have enabled the east at last to rejoin the west. For their part western Marxist and neo-Marxist accounts, which have had problems with grappling with 'actually existing socialism', have tended to see the outcome of the Cold War not so much as a victory for capitalism but as a defeat for authoritarian state socialism (Denitsch, 1990). Alternative 'third ways' to embracing either neo-classical market economies or Stalinist-type development do exist. Callinicos (1991:2) takes this argument further in suggesting that 'what is dying in the disintegrating Eastern bloc is not socialism, of however a degenerate and distorted form, but the negation of socialism'. But however utopian such ideas seem at present – of post-state socialist societies turning away from their universal embracement throughout the region of capitalism – there is little doubt that the collapse of 'actually existing socialism' may not follow the smooth path towards full membership of the capitalist world-economy and of a global community of democratic states that some theorists suggest. What is certain is that the collapse of state socialism carries profound implications for the geographical world in which we live and our interpretations of it.

Both the relative merits of models of hypothetical social systems, and the understanding of a real and substantive historical and geopolitical struggle between the two worlds of capitalism and socialism, are well-established areas of intellectual concern (for a review of the literature, see Rutland, 1989). What is more problematic is that so far we do not possess appropriate alternative theories or models to understand half the world's transition from state socialism to an as yet uncharted post-socialist order. Yet all this seems strangely unfamiliar terrain to geographers. While cognate disciplines have been discussing and debating the causes and implications of such momentous events, geography has been embarrassingly silent. As Folke and Sayer note with hindsight, 'it was extraordinary that the crisis of socialism was rarely confronted directly' (1991: 240). It would seem that for too long geographers either ignored the socialist world or treated it as tangential, important only in so far as it has had a bearing upon capitalism (see, most notably, Harvey, 1989; Johnston and Taylor, 1989; Kobayashi and Mackenzie, 1989; Macmillan, 1989; Peet and Thrift, 1989). For their part, Area Studies specialists have tended to focus on the spatial impacts and consequences of economic reform, paying little or no attention to how political and social theory might advance our geographical insights.

This chapter is a contribution to rectifying this omission from our geographical agenda. It addresses itself to the end of a particular form of geopolitics, that of the Cold War, and to the consequences that its ending has had for the reconstitution of the socialist world. It argues that a complex choreography existed to the geopolitics of the Cold War, which first needs to be convincingly mapped out if we are to understand its end and also that of state socialism. In turning to examine the transition from state socialism, it suggests that any understanding of the geopolitical reconstitution of that world must acknowledge the ideological ascendancy in post-state socialist societies of both capitalism and nationalism, which I examine in relation to the geopolitical transition now under way in the one-time regional epicentre of the Cold War, Europe.

THE GEOPOLITICS OF CAPITALISM AND SOCIALISM: STANDARD INTERPRETATIONS

Most approaches to the post-1945 world order fall short of explaining the Cold War's uniqueness, durability and ending. This is characteristic of three approaches in particular.

1 Traditional geopolitics This approach has a long association with geography (Parker, 1985). Utilizing in particular (although not exclusively) the works of Mackinder and Spykman, traditional geopolitics drew upon such notions as the heartland–rimland concept, and on geopolitical concerns with the domination of Eurasia by one superpower, to construct an explicit theory of the Cold War. Its basic, western-centric assumption was that what gave the Cold War its underlying momentum was Soviet expansionism (for example, Gray, 1977). It was the USSR's spatial aggressiveness, causally linked in some discourses to the dominant political ideology and/or to particular historical and geographical features embedded in Russian/Soviet society, which threatened world peace. As Dalby (1990) illustrates, it was a world view which underpinned American Cold War security discourse and was translated into a policy of western encirclement, as reflected in the idea of geostrategic containment and the domino theory. By portraying the Soviet Union as *the* expansionist power, driven by an ultimate goal of world domination, the United States could legitimize military intervention in a number of regional hotspots, in the name of protecting 'the free world' against further Soviet encroachment.

2 Balance of power models According to this line of reasoning, embedded in particular in the realist literature on international relations, the Cold War was maintained through globally and regionally organized state hierarchies of power, defined in relation to the means and capabilities by which states protect their national interests. This involved focusing in particular on those states at the apex of the geopolitical world order, the United States and the Soviet Union, whose superpower credentials and military and geopolitical parity were taken as fundamental to maintaining geopolitical equilibrium (Cohen, 1982; Jonsson, 1984).

Emphasis lay on understanding order, hierarchy and stability. But in viewing world politics solely through the prism of an inter-state system fluctuating around a geopolitical equilibrium, such accounts failed to predict the Cold War's eclipse. Moreover, by focusing more or less exclusively on inter-state relations, both non-state global actors and socio-political changes occurring within domestic state politics and society tended to be treated at best as contingent developments. Realist accounts thus failed to grasp fully the dynamism and complexity of the post-1945 world order (Linklater, 1990).

3 World systems theory In contrast, by taking as its frame of reference the world capitalist economy and interpreting the state and its external relations as the political organization of the world-economy, world systems theory moves away from the state-centrism of realist accounts. For Wallerstein (1984), the modern world embraces one mode of production, founded on market exchange, which determines the nature of inter-state relations based on a state's location within the world-economy. Such an approach has been subject to a critique of economic reductionism. 'The non-reducibility of state institutions to mere "superstructures" ', wrote Frankel, 'precludes the conception of the world as simply a capitalist world' (1983:179). In other words, the complexity of bilateral and multilateral political, military and other relations explodes the myth of a single 'core', 'semi-periphery' and 'periphery'. In following a similar line of critical reasoning, Worsley (1984) argues that any framework for understanding the global political economy must recognize not only its increasingly interdependent nature but the existence of another polarity which dominated most of the post-1945 world order, that between the two competing social systems of capitalism and Communism.

This does not, however, mean endorsing Syzmanski's (1982) interpretation of the two world systems as economically independent of one another. As Chase-Dunn (1989) and other world systems theorists rightly note, even at its geopolitical apogee what was portrayed as 'a socialist world-economy' was not economically autonomous and self-sufficient. But this, in turn, does not mean labelling socialist states as 'state capitalist' by virtue of their economic dependency in a world economy in which capitalism holds sway. To ignore the differences between these two systems is to erase the differing ways in which socialist countries organized their economies, and to suppress the distinctive features which governed their political, social and cultural lives (Smith, 1989a). Moreover, the assumption of a world in which anti-systemic forces, embedded in the constitutive nature of capitalism, provide the social motor for constructing a socialist alternative is contrary to what contributed to the downfall of state socialism in Europe, namely, the role of social forces antithetical not to capitalism and its functioning (as Taylor, 1991a, argues), but to 'actually existing socialism'.

AN ALTERNATIVE THESIS: POST-STALINISM AND ATLANTICISM

Reconstructing the Cold War can therefore begin by acknowledging that what made it unique in the history of geopolitical relations between states was that it constituted a conflict between two rival social systems, capitalism and socialism. The clearest exposition of this dimension to the Cold War is provided by Halliday (1983, 1990). According to this account, what underpinned the post-war global system was its inter-systemic character, as reflected in three aspects of the systems' inter-relationships. First, each was organized on the basis of contrasting social principles, one based on private ownership of the means of production, the other on state ownership. State socialist control over social production and distribution was exercised by a highly centralized state apparatus which monopolized political, economic and ideological power. It is a difference which also carried important implications for the nature of domestic politics and for the degree of global influence and flexibility that leading capitalist states possessed over their socialist counterparts. Through, in particular, the medium of such economic institutions as the International Monetary Fund and multinational companies, western capitalist states (especially the United States) possessed an unrivalled role in the world economy.

Second, both systems had an ideological claim to be world systems; that is, ideal societies which others should follow. Thus what made the era of the Cold War unique was not state ideologies of 'national exceptionalism' *per se* (Agnew, 1983) but rather those of 'systemic exceptionalism'. Each possessed a belief in moral separateness and the superiority of its system over the other. This was also reflected in the systems' Cold War geopolitical discourses, and played a part in structuring policies of national security. Thus the Truman doctrine, which involved the 'containment of an ideology' (Steel, 1977:22), became for the United States a global doctrine legitimizing US intervention in the name of pre-empting Soviet expansionism, while in the same way the Soviet Union's 'two-camp thesis' – of a purported ongoing struggle between socialist and imperialist worlds – helped to structure Soviet third world policy. The goal, then, was the prevalence of one ideology over the other. As E. P. Thompson (1985:44) wrote:

> It is ideology, even more than military-industrial pressures, which is the driving motor of the Cold War ... It is as if ... ideology has broken free from the existential socio-economic matrix within which it was nurtured and is no longer subject to any control of rational self-interest. Cold War II is a replay of Cold War I, but this time as deadly farce; the content of real interest-conflict is low but the content of ideological rancour and 'face' is dangerously high.

Finally, inter-systemic conflict had a tendency to override attempts at state- to-state accommodation, including superpower accommodation. This had particular implications for the third world. As Halliday (1983: 33) noted, 'the very social interests embodied in the leading capitalist and communist states are present, in a fluid and conflicting manner, in third world countries; the result is that the clash of the two blocs is constantly reanimated and sustained by developments in these other states that may be supporters and allies of one or other bloc'. This was of course reinforced by superpower aid and arms sales, which tended to follow a spatial logic reflecting respective superpower geopolitical interests and mutually accepted regional spheres of influence. In all, such tendencies furnished a predictability to global geopolitics.

It was precisely these particularistic ideologies, along with superpower nuclear capability, which made the Cold War geohistorically unique. Yet to highlight the underlying differences which defined the nature of the two systems does not mean simply reducing the Cold War to what Deutscher (1960) called 'the great contest'. Neither capitalism not socialism existed in unadulterated form. In other words, recognition of the importance of inter-systemic differences should not lead us to portray each system as homogeneous, with the two core states simply imposing their particular social systems upon their respective worlds. Varieties existed in each (compare the post-war Keynesian British welfare state with the United States, and post-1956 Hungary with the Soviet Union). Rather, it is more useful to see the Cold War as constituted on the basis of two social and geopolitical systems – Atlanticism and post-Stalinism. As social systems, argues Kaldor (1991:35), both 'became archetypical because all other variants were excluded'. Moreover, 'these two systems ... were not in conflict but were *complementary*, tied together by the same historical experience. Each needed the other. Both required high levels of military spending and a permanent external threat. The existence of each provided a legitimation for the other' (ibid.). But what is also crucial to explaining the Cold War is the differences that distinguished post-Stalinism from Atlanticism. Three such can be identified: geopolitical formations, relationships between geopolitics and militarism, and the capacity to economically restructure.

Geopolitical Formations

Both post-Stalinism and Atlanticism were born out of the Second World War. The former constituted a particular phase of socialist history which, in its geographical reach, was the product of the forcible export of 'socialism in one country' to 'socialism in one region'. Eastern Europe was in effect reorganized to fit the template of Stalinism. There were three core elements to this regional hegemonic system: the imposition of a single socio-political model (totalitarianism); the denial of independent, sovereign and equal status to the East European states; and the establishment of a centrally imposed stratagem of divide and rule, as reflected in the *de facto* bilateral structure of the bloc's two most important regional organizations, Comecon and the Warsaw Pact (Dawisha, 1988).

With de-Stalinization, a greater degree of geopolitical flexibility and state autonomy was permitted and the principle of countries pursuing 'different paths to socialism' was acknowledged, but only in so far as this did not compromise the regional hegemonic power. In maintaining Soviet rule, the role of local Communist parties as instruments of central control remained paramount to the continuation of the post-Stalinist system; the principle of socialist internationalism, which re-emerged in 1968 following the Czech crisis in the form of the Brezhnev doctrine of limited territorical sovereignty, reflected the imposed subordination of, and the geopolitical limits to, putting national interests before those of (Soviet) bloc socialism (Jones, 1990). Such a regional hegemonic system also ensured client dependency in the economic sphere by, in particular, imposing geopolitical limits to introducing structural change. East European states thus remained structurally dependent on Soviet raw material and energy supplies, while in imposing geopolitical limits to reform Moscow made it near impossible for its client states to compete effectively in global markets. In short, post-Stalinism constituted a geopolitical and economic system of tight superstate control and stultifying rigidity.

What underpinned Atlanticism was an economic system of international capitalism, centred on the United States and its transnational corporations in alliance with Western Europe. The system's economic stabilization and functioning was ensured through a series of geopolitical alliances stretching arc-like from the Atlantic (NATO) to South East Asia (SEATO), and incorporating a series of third world states as part of a western-centred global trading system. Although possessing its own regional hegemon, the United States, which provided both monetary and military leadership, compared with post-Stalinism, Atlanticism resembled more of an alliance between sovereign states. It was of course not without its geopolitical discord, as reflected in such episodes as France's departure from NATO, West Germany's *Ostpolitik*, and waning West European support for Atlanticism following US regional interventions in the Middle East and Central America. Atlantic Europe's noticeably more accommodative stance towards the post-Stalinist system, particularly apparent during the years of the early 1970s' detente and in part linked to the greater benefits which had accrued to the region in the economic and security fields (Malcolm, 1989:35–6), was also a source of trans-Atlantic tension, exploited by the Brezhnev regime in its later years in a regional appeal to the idea of a common European home.

What, however, cemented the centrepiece of Atlanticism – the United States and Western Europe – was their shared economic and geopolitical interests, reflected in what Kaldor (1991) calls 'the Atlanticist compromise'. Of fundamental significance was the role that anti-Communism played as a way of persuading even hardened US isolationists that the United States should play a pivotal economic role in Western Europe. This point is most forcibly made by Calleo (1987:196), who argues that even in the economic sphere 'it took the Cold War to mobilise American domestic support for the massive aid Europe required to participate in a trans-Atlantic liberal system. Had no Russian military power pressed them, Europeans would have resisted America's economic penetration, even more than they have.' Marshall Aid, as George Kennan noted, had two primary objectives, 'so as they can buy from us' and 'so that they will have enough self-confidence to withstand outside pressures' (cited in Barnett, 1983:114). Thus the Marshall Plan was a type of *acte gratuite* in economic terms, based on the sound geopolitical principle that greater security would result from having prosperous allies. And the Cold War itself, through the geomilitary structure of NATO, helped to facilitate both a continuous transfer of resources and the spatial diffusion of Fordist methods of production through the creation of a trans-Atlantic infrastructure.

The anti-Communist stance of periodic socialist governments in Western Europe, notably in Britain, Italy, France and the Netherlands, was a way of making socialism and the idea of the interventionist state as part of Atlanticism palatable even to conservatives within the United

States. With the emergence of the New Right in the early 1980s on both sides of the Atlantic, Atlanticism entered a new phase of heightened anti-Communism which strengthened intra-bloc discipline. It was anti-Communism which in part justified massive increases in military spending, and tolerance within the United States of a growing US deficit, linked to the Reagan administration's geopolitical ambitions of curtailing the activities of 'the evil empire'.

Geopolitics and Militarism

Geopolitics was central to giving militarism a different character in the post-Stalinist bloc to that in its Atlanticist counterpart. For our purposes, militarism can be broadly defined as 'a set of attitudes and social practices which regards war and the preparation for war as a normal and desirable social activity' (Mann, 1988:166). Two types of militarism identified by Mann are particularly useful: 'militarized socialism' and 'spectator militarism', which can be usefully applied to post-Stalinism and Atlanticism.

From the outset of its formation, the Soviet Union was conscious of its geostrategic insecurity in a world of nation-states in which capitalism holds sway. Indeed, the developmental choices open to the Bolsheviks after the 1917 Revolution reflected a growing recognition, as the following decade progressed, that securing the aims of 1917 and thus the economic transformation of an overwhelmingly rural and backward society was bound up with the need to be militarily strong and economically independent from the capitalist world economy (Smith, 1989b). The eventual developmental path chosen, 'socialism in one country', was in effect a defensive strategem structured and legitimized to ensure rapid military-industrialization, an outcome which can also be viewed as bound up with centuries of Russian geopolitical insecurity and recognition of the causal relationship between successive military defeats and the country's technological and economic backwardness. In Lang's (1962) terms, what was created by Stalin was a war economy, which sacrificed both balanced development and local democratic accountability for the military-industrial complex and the strong state. It was this war economy which was to provide the structural basis for the post-Stalin economies of both the Soviet Union and its post-1945 client states. In short, the ease with which state socialist economies could adjust to the Cold War reflected the very militarist structure upon which their economies had been built and in which they best functioned. But the centrality of the military-industrial complex was also to become post-Stalinism's Achilles' heel, for it ensured both the deprioritization of investment in social needs and also imbalanced sectoral development.

The significance of the military-industrial complex was not confined to economic life. Militarism pervaded Soviet society. In particular, the so-called 'Great Patriotic War' played a crucial part in the making of what the Soviets themselves referred to as 'the military-patriotic society' and of ensuring that socialism, once and for all, became a piece of the Soviet national heritage (Hall, 1985). Various war-time tools for social mobilization, tailored to meet geopolitical needs, were simply refined to fulfil the requirements of the Cold War. These pervaded the organizational structure of more or less all social, economic and political institutions, from collective farms to factories and from trade unions to schools. Being a good citizen in whatever walk of social life was being responsive to military-industrial discipline. This was reinforced by military conscription, which structured 'the political consciousness ... by forming two conditional social reflexes: the reflex of unquestioned submission to authority and the reflex of automatic dichotemisation of the world into "ours" and "theirs"' (Zaslavsky, 1982:30).

Whereas militarized socialism institutionalized global militarism in social life, so providing an important motor for social stability, the same was not the case with Atlanticism. This is summed up in the metaphor 'spectator militarism', in which society was not so directly implicated in the Cold War. Of course militarism did have a major impact on both sides of the Atlantic in structuring labour markets and on industrial and spatial development (Lovering, 1987), as well as in creating a common Atlanticist Cold War culture through the medium of film and television. Yet in comparison with life under post-Stalinism, national citizens were mobilized not so directly as players but as spectators, a fact linked in part to the more peripheral and less institutionally pervasive role that militarism played in western life. For post-Vietnam US society in particular, geopolitical intervention was more distant – the Cold War was often fought out by a client state. Popular support for military intervention was also politically more volatile, linked to public debate concerning the costs and benefits of military as opposed to social welfare and other forms of governmental spending.

Differences did, however, exist between the two halves of Atlanticism. In Atlantic Europe, the direct geohistorical experience of war and location at the geopolitical epicentre of the Cold War gave militarism a far higher social profile, and indeed was pivotal in the eventual emergence of organized 'grassroots' opposition in the form of the non-aligned peace movements. These were not only active in challenging the Cold

War's continuation but, in connecting up with similar movements in Eastern Europe, gave a particular European focus to concern over 'outsider' superstate militarism. For these movements, the question became one of security for whom and at what environmental and social cost.

Reform and Stagnation

By the early 1980s, many commentators began to see in both economic systems and in their respective superpower states the beginnings of economic decline, if not of systemic crisis. Analysis focused in particular on factors involved in the relative decline of superpower state hegemony within their respective systems (for example, Keohane, 1984; Konrad, 1984; Start, 1985; Bialer, 1986; Calleo, 1987; Cox, 1987; Sokoloff, 1987; Gill and Law, 1988; Kennedy, 1989; Pugh and Williams, 1990). It was, however, the United States which received most attention.

For many the apogee of US economic and geopolitical hegemony was past; studies highlighted the United States' decline in share of the world's wealth in industrial production (falling from over half in the early 1950s to a fifth by the early 1980s) and to an increasing budget deficit, linked to its scale of military commitment to Atlantic Europe and to large expenditures on increasingly sophisticated military hardware. In addition, there were new challenges by Germany and Japan in the global market-place – two economic superpowers whose meteoric rise, it was contended, was a product of their comparatively low levels of gross national product (GNP) devoted to the Cold War and of the consequently larger levels of investment earmarked for civil industry and new technologies. This decline was most succinctly formulated in Kennedy's (1989) highly influential thesis that expanding US global interests and obligations, like those of previous hegemonic powers in the seventeenth (Holland) and nineteenth (Britain) centuries, had outstripped economic generation at home.

There seemed little doubt that decades of geopolitical intervention elsewhere were having serious economic and psychological consequences for the United States' role in the capitalist world-economy. Where disagreement tended to occur was primarily over the extent of decline, although Strange (1987) cautions against those who refer to absolute decline in what she regards as the myth of lost hegemony. For her it is important that any examination of the United States considers its unique influence and capacity to continue to shape the structures and institutions of the global political economy. Thus US predominance is seen as based on four main areas of what she labels 'structural power': security (the ability to protect other states against external threat); production (who shall produce what, how and with what reward); financial (the ability to control the supply and availability of credit); and knowledge (the facility to control both the channels by which knowledge is communicated, and access to them). Much of this seems particularly sanguine in a post-Cold War era in which, in terms of global security at least, unipolarity is as apt a description of the geopolitical order as any other.

What is more relevant, however, is not so much the extent of the rolling back of US hegemony or the way in which its global role was being reconstituted, but that two important economic changes were occurring to Atlanticism during the 1970s and 1980s, both of which were to have consequences for post-Stalinism.

The first concerned the increasing globalization of capital, driven largely by the core economic states and global economic institutions of Atlanticism. Both trade and investment had become increasingly internationalized; large increases in production could be realized by organizing production across state borders, a process which was also matched by enormous international flows of financial investment (Agnew and Corbridge, 1989; Thrift, 1989a). This global integration of capital which transcended geopolitical divisions between capitalist countries did, however, have its geopolitical limits, for it left the post-Stalinist bloc largely excluded.

Secondly, those structural changes to the capitalist world economy were accompanied by the introduction of new forms of production and patterns of consumption; in short, the rigidities of Fordism, which had dominated Atlanticism throughout the early Cold War years, were being replaced by post-Fordist practices of flexible accumulation. Such changes represent nothing short of a qualitatively new phase of capitalist development (Harvey, 1989). This dynamism to capitalism, irrespective of the highly differentiated geographies and new patterns of development which it has produced, reproduces and is recreating, represented state socialism's greatest failure – the underestimation of capitalism.

This dynamism also contrasted sharply with the stasis which became an increasing feature of the post-Stalin system, despite periodic attempts, particularly during the Khrushchev administration, to wrestle with economic reform. Experimentation with political change, however, was regarded as a more direct threat to Soviet hegemony in both its internal and external empires, and was avoided. Strict geopolitical limits to reform were therefore applied. In Eastern Europe, where economic reform went furthest, uncoupling any such attempts from political reform was crucial to avoiding a geopolitical backlash and thus justification for Soviet military intervention. Thus Hungary after 1956 learned that, in abandoning political pluralism and thus not challenging Soviet rule, it could embark upon the limited introduction of the market into the planning system. In failing to uncouple political from economic reforms, the Czechoslovaks in 1968 were less fortunate.

By the late 1960s, the Soviet regime had more or less abandoned reform for stability, both at home and abroad. Despite the benefits which might have accrued to labour productivity and living standards as a result of economic liberalization and technocratic reform, the price of overhauling Stalinism was judged as carrying too great a social and political cost. As Bialer (1986) notes, economic reform would have involved temporary material sacrifices for the citizenry, as well as setting in motion institutional changes which would have threatened the privileges that the bureaucratic middle stratum and Party *apparat* had come to enjoy. What in effect had been created was a Stalin Mausoleum (Z, 1990) unable to effect economic and political change. In part, it was linked to an economy which, in the absence of war, had replaced regulation of battle by bureaucratic regulation, but which could not escape from a central planning system whose targets were geared towards the geopolitics of war, and which could not effectively reajust to internal needs (Kaldor, 1990:31).

In many respects, Kennedy's (1989) thesis of 'imperial overstretch' seems more appropriate to the Soviet superpower. The absence of economic reform affected not only productivity but the capacity of the Soviet Union to finance both its empire and the increasingly expensive technology and other investments associated with the Cold War. Finding resources to manage its external empire was becoming ever more difficult. As Bunce (1985) notes, whereas in certain respects (primarily in terms of security), Soviet–East European relations can be considered in the first 15 years of post-Stalinism as 'Soviet assets', in the last 15 years they were being transformed into 'Soviet liabilities'. This was a result of: first, the economic and political dependence of Eastern Europe on the core economy; second, the costs to the Soviet Union of maintaining its East European monopoly of political and, to a lesser extent, economic power; and third, the costs which accompanied the bloc's limited reintegration with a global capitalism that was undergoing an economic downswing. Whereas increases in costs of energy from 1973 onwards in the world-economy strengthened Soviet hegemony over Eastern Europe, falling oil prices on the world market in the mid-1980s had the opposite effect. Unable to provide the resources necessary to stimulate economic growth in Eastern Europe or to permit reform, the East European countries had increasingly to finance their own modernizations by looking westwards, a process which included borrowing heavily from western economic and political institutions.

Internally, too, there were considerable social costs. As Zaslavsky (1982) notes, during the 1970s and early 1980s (the so called era of 'developed socialism'), the Soviet state did not simply rely on coercion to ensure internal social stability but rather also maintained its hegemony by exercising social benevolence. This took the form of a series of centrally imposed social contracts in which, in return for limited access to political decision-making, society was in effect treated to a number of benefits, including full employment, a welfare state and a slow but steady rise in living standards. Yet two important changes were undermining this 'organized consensus'.

Firstly, society was undergoing a quiet and unnoticed revolution in the 1960s, embedded in what Lewin (1988) calls 'the urbanisation of society'. In 1960, the urban population was barely 50 per cent of the total; by 1985 it had risen to 65 per cent (and over 70 per cent in the European Soviet republics). This accompanied important changes in the social structure and, in particular, the rise of a new hegemonic social formation of diploma holders ('intelligentsia' in Soviet parlance) who, in the increasingly diploma-oriented society of 'developed socialism', came to dominate not only the culturally related professions but also other public spheres of economic and political life. It was people from the ranks of this disparate urban social stratum, 'the Gorbachev generation', who came to challenge those in the previous dominant social formation, who had gained position primarily as a result of loyalty to the Party. It was a more cosmopolitan-minded and educated elite, frustrated by the whole project of post-Stalinism: it was made up of, for example, engineers and scientists whose professionalism had become strait-jacketed by a production system contrary to economic growth and who displayed widespread job dissatisfaction, and of those in the cultural professions whose employment was a daily reminder of the state-imposed limits to basic civil and political citizenship.

Secondly, by the late 1970s and early 1980s another aspect of this 'organized consensus' was coming to an end: living standards could no longer keep up with the rising consumer expectations of this more urbanized and educated society. Differentials in living standards, for instance, were re-emerging. Regional policy, which had been designed to bring the living standards of the more backward regions up to the level of the more urbanized, was no longer as effective, with the poorer republics of Central Asia falling behind the Baltic republics, Russia and the Ukraine (Smith, 1991). In Eastern Europe, standards of living were also beginning to fall dramatically, in part due to the introduction of austerity measures linked to having to repay western debts. Militarism and the absence of reform were undermining support for both the social and the geopolitical status quo.

What put post-Stalinism, including its inability both to redress economic stagnation (*zastoi*) and to guarantee improved living standards throughout its domain, in an especially bad light was the comparative geopolitical context. This, in particular, meant the record of post-Stalinism relative to that of Atlanticism. This carried profound implications for the legitimacy of state socialism. The contrast of this record with the belief in the inevitable crisis and decline of capitalism, and in the ability of Communism to overtake it and constitute an alternative rival, independent of the capitalist world, exposed the whole state-projected image of socialism as catching up and overtaking capitalism. The breakthrough into modernity had eluded post-Stalinism. Moreover, the dynamics of capitalism contained another new dimension to its global impact; an increasing role came to be taken by forms of activity associated with communications and consumer culture, from which the post-Stalinist bloc could not insulate itself in spite of decades of attempts to maintain a *cordon sanitaire* against capitalism (Halliday, 1990). Media technology, for example, provided an electronic window through which the peoples of East Berlin (in terms of West German TV) and of Tallinn (via Finnish TV) could compare the reality of life under state socialism with the projected image of capitalism.

While such preconditions were necessary to ensure the beginning of the virtual end of post-Stalinism, what separated the revolutions in Eastern Europe from that in the Soviet Union was that, in the home of the Bolshevik Revolution itself, it was psychologically and politically impossible to abandon state socialism overnight (Gellner, 1990). In part, this is reflected in what Habermas (1990) calls 'the rectifying revolutions' in Eastern Europe, where, following the withdrawal of the Brezhnev doctrine, one finds the popular desire to *reconnect up* with the inheritance of constitutional democracy and, socially and politically, with the styles of commerce and life associated with capitalism. In the Soviet Union, post-Stalinism was considered as reformable, as was captured in the notion of *perestroika*, and it was believed that a Leninist 'middle way', drawing in particular upon the mixed space economy model of the pre-Stalinist New Economic Policy, could provide the material base for economic prosperity. But this 'fig leaf capitalism' was unable to resolve the economic stagnation that reform Communism inherited (Sakwa, 1990). The failure of reform Communism to improve society's economic well-being also went hand in hand with increased demands, by social groups throughout the localities, to secure far greater local democratization than *perestroika* was able to deliver.

GEOPOLITICAL TRANSITIONS

As geopolitical formations whose very *raison d'être* depended on the Cold War, post-Stalinism and Atlanticism are no longer central to understanding the geopolitical restructuring of post-Cold War Europe (Lee, 1990; Taylor, 1991b). Neither the economic (Comecon) nor military (Warsaw Pact) organizational structures which defined the parameters of post-Stalinist hegemony in Europe have survived the post-Cold War era. Eastern Europe has in effect become detached, both economically and geopolitically, from its eastern neighbour. Atlanticism has also been undermined. NATO has undergone re-evaluation in both halves of Atlanticism. For its part, the United States is re-evaluating its role and the costs both of remaining part of a European security structure and of the challenges which a stronger Europe represents for not only regional but also global US markets. Are we then to assume that the two halves of Europe are on a trajectory to closer integration?

Imposing transition models on societies undergoing geographical restructuring is a perilous exercise (Thrift, 1989b). We know where the two halves of Europe have come from (although the exact labelling can be debated), but it is too early to know what eventual form and shape they will take, let alone declare that 'history is ending'. It is, after all, muddled transitions that dominate in history and not the ideal types around which strong theories are built (Chirot, 1985). It is far safer to consider the normative geographical imaginations of what is Europe and how they square with the reality of differential development, geopolitics and cultural traditions. Three such visions can be identified and briefly explored.

Common European Home

In its most current form this conception of Europe 'from the Atlantic to the Urals', reflecting a Gaullist vision of Europe aimed at reducing the

United States' influence, is associated particularly with the project of reform Communism and its main architect, Mikhail Gorbachev. In his Prague speech, Gorbachev envisaged Europe as 'an historical and cultural category . . . in a spiritual sense' appealing to 'the peaceful development of a European culture, which has many faces, yet forms a single entity' (*Pravda*, 11 April 1987). It is a project that has been open to a number of interpretations: both as a geopolitical stratagem to drive a wedge between the two halves of Atlanticism and to secure Soviet influence, and as an economic strategy to enable the investment and technological impulses needed from Western Europe to facilitate the success of reform Communism (Waever et al., 1989). In the post-Cold War era, however, rather than using this as a strategy to exclude US influence, the Soviet leadership recognized that both superpowers should be included in any new European security structure. 'The USSR and United States', declared Gorbachev in his landmark 1989 speech to the Council of Europe, 'are a natural part of the European international political structure. And their participation in its evolution is not only justified, but historically conditioned' (*Pravda*, 7 July 1989). German reunification in particular has been crucial to such rethinking. Indeed, for the USSR, agreeing to the continuation of a US-backed presence through NATO to include a united Germany was one way of providing an institutional constraint for what Moscow feared could become an unrestrained major power in Europe (Stent, 1991:151).

Yet becoming part of a common European home did not square entirely with the Soviet Union's geopolitical self-image. As Hauner (1990) reminds us, the USSR was a Eurasian power, and so the idea of a common European home must also be placed in the context of Gorbachev's claims to 'the Pacific, our common home' and 'the sub-Arctic, our common home'. It is also a legacy which its largest successor state, Russia, has inherited. This not only carries implications for the nature of its integrative orientation into the world economy, but also has geopolitical ramifications linked to Russia disentangling itself from its former status as the core state of a multiethnic empire which bridged two continents. Consequently, geopolitical relations with the other former European and central Asian republics are likely to be problematic, not least due to the uncertain future facing Russian minorities in these non-Russian states. Geopolitical problems, in short, integral to nationalist tensions, are unlikely to secure for Russia the role of *primus inter pares* in a new economic community of post-Soviet states stretching from the Baltic to the Pacific.

'Not Quite All Europe'

Defining the geopolitical parameters of a 'not quite all Europe' centred on the European Community is a precondition to this becoming part of its project for a deepening and more inclusionary framework for Europe, based on the idea of liberal democracy and the market. In terms of the former, it reflects a European strand of thought which sees the whole project begun by the Enlightenment and the French Revolution as completed in Europe by the 1989 revolutions; in short, as a victory for western-based ideals – a conception of Europe reflected in Bender's (1981) idea of Europeanization emerging out of 'the end of ideology'. For some Europeans, however, embracing the market does not necessarily mean adopting the unfettered free-market capitalism of the New Right, but rather of a community offering an alternative 'third way', based on the socially regulated capitalist models provided by Germany, Italy and Sweden. It is a model which emphasizes the importance of economic coordination, regional regulation, and cooperation between industry, labour and the state in order to ensure both economic prosperity and distributive justice (for example, Hirst, 1989).

Much of the power to shape this 'not quite all Europe' is based on the institutions born out of Atlanticism (the EC along with the International Monetary Fund, the World Bank, COCOM, etc). In part it is captured in Pinder's (1991:98) question as to whether the omission of Eastern European states from the original Marshall Plan 'leaves unfinished business in the form of its completion . . . which would lead to their integration into the European Community; and that, like the original Marshall Plan, this would respond to the connection between economic need and political danger'. Substantial differences exist, however. First, the United States is no longer the central actor and donor, having been usurped by the European Community and its member-states. And second, post-1945 Western Europe could have been expected to recover and reintegrate into the world market economy relatively quickly after six years of war, whereas four decades of state socialism will take Eastern Europe much longer to recover from. But like the original Plan, it is not a giveaway programme undertaken for non-altruistic economic and geopolitical reasons. Conditionality has become a precondition both to full membership and to receiving aid for restructuring and access to Community markets.

Typical is the Community's Programme for Eastern Europe's Economic Restructuring (Phare Programme), which began in 1990 by providing aid to Poland and Hungary, and then was extended in the following year to the rest of Eastern Europe. Not only must aid benefit the private sector in particular, but recipients are required to switch a proportion of their own budgetary resources towards similar projects that would also benefit the private sector. For Gowan (1990), this reflects a privatizing mission by western economic and political institutions and a direct intervention by the west in internal East European debates as to how their national economies are shaped. Much, though, will also depend upon how Eastern European states respond to the needs of capital accumulation after decades of intra-bloc protectionism and non-market competitiveness. Comparisons have already been drawn with Spain and South Korea (Hankiss, 1990). While the former succeeded in combining its transition from authoritarianism to democracy with capitalism and integration into new global and regional markets, the economic success of the second was at the price of an increasingly authoritarian, *dirigiste* political system. The fear here is that the need to attract western investment capital to ensure economic growth will demand not only high social costs but also the re-emergence of a coercive state, having to forfeit some of the democratic gains of 1989.

A Europe Back to the Future

This conception of Europe is a salutary reminder that the most striking feature of the rebirth of civil society in the east is the salience of ethno-regional divisions, and of the power of some nationalist movements to successfully fulfil the objectives of a political ideology rooted in a tradition of European political thought and given concrete meaning in the twentieth century in the Wilsonian idea of natonal self-determination. This challenge to Europe's geopolitical map is, in a curious way, returning the region to the problems associated not with a geopolitical settlement drawn up at Yalta in February 1945, but rather with a European map unsatisfactorily formulated between 1918 and 1922.

This post-World War I map, a product of the Versailles Peace Settlement and the Russian Revolution, was based on two alternative but geopolitically flawed visions of national self-determination, one Wilsonian and the other Leninist. Both were geopolitical programmes designed to provide stability in the ethnic patchwork of territories ruled over by the Austro-Hungarian, Ottoman and Russian empires. As Hobsbawm notes, the Wilsonian principle proved 'utterly impracticable' because 'it attempted to make state frontiers coincide with the frontiers of nationality and language' (1990:132). It merely transformed the status of a large number of 'unsatisfied nations' (such as the Slovenes, Croats and Macedonians) from that of 'oppressed peoples' to that of 'oppressed minorities' in new nation-states constructed around superordinate ethnic groups, some periodically if not consistently hegemonic in their form of governance (for example, Serbian-dominated Yugoslavia and Polish-dominated Poland). Indeed, the logical implication of Wilsonian politics, of attempting to manufacture a coextensive relationship between territory and nationality, was to involve and to legitimize the mass expulsion of minorities. It was only when the Great Powers of the inter-war years intervened that the Wilsonian logic of attempting to match sovereign territory with ethnicity and religion was breached, as in the case of Germany's claims to part of the German-speaking eastlands and also as a result of Hitler's accommodation with Stalin concerning 'regional spheres of influence'. (As a result the Baltic states whose independence had been established in 1918, and Romania's eastern territory now known as Moldova, were incorporated into the Soviet Union.)

Further east, the Leninist stratagem of national self-determination was to endure much longer. This was because the Soviet state successfully internalized its nationalities' problems, legitimizing the creation of a nationality-based Soviet federation in 1922 on the claim of simultaneously forwarding 'proletarian internationalism' and the right of nations to cultural and territorial-administrative autonomy. It did not alter the fact, however, that the new multicultural polity was largely a product of a deal struck between Moscow and provincial Bolsheviks, whose support for a Soviet federation was mostly confined to the major cities of the non-Russian lands. It was to become a form of rule which, into modern times, required at least two things to maintain territorial stability: highly centralized political power backed up by an acceptance of the state's willingness to use force; and an adoption of social benevolence in such spheres as regional policy (particularly in relation to improving the living standards of the less-developed republics) and the affirmative action policies desired to assure the upward mobility of natives in their homelands (Smith, 1991). Both *glasnost'* and democratization, however, challenged and exposed the centre's myth of a multi-ethnic periphery identifying and belonging to a 'new historical community ... that of the Soviet people [*Sovetskii narod*]' based on 'sharing a common territory, state, economic system, culture, the goal of building communism and a common language' (*Materialy XXIV se"zda KPSS*, 1971:76). Post-Stalinism had in effect been unable to realize a Leninist vision of erasing nationality differences, or to provide basic local freedoms and living standards comparable to those of western capitalist society.

At the forefront of championing the demise of the Soviet federation has been the local cultural intelligentsia. These traditional bearers of nationalism have led their republic-based popular fronts from initially supporting reform Communism's 'revolution from above' to orchestrating the end of the Soviet empire. Their ability and the speed with which they could mobilize their constituents behind the separatist cause did, however, vary. At the cutting edge were the Baltic republics, whose demands to rejoin the European community of nation-states reflected (to paraphrase Anderson, 1983), peoples whose sense of nationness had long since been 'imagined as sovereign'. Such imaginings are not restricted to those Soviet republics that can draw upon the powerful symbol of pre-Soviet nation-statehood, or indeed to only one socialist federation. Both Yugoslavia and Czechoslovakia have been unable to prevent the end of their multi-ethnic polities. It would seem that history has proved Lenin right in viewing the socialist federation as a transitional form, but not in the way in which he envisaged.

There is thus a paradox at bay. On the one hand, the transition to capitalism requires integration and membership of a larger Europe in order to ensure the east's economic modernity. Yet on the other hand, nationalism wishes to reconstruct the more fragmented and insular world of a past era. Indeed, for the first time since the political fragmentation of the empires of Central and Eastern Europe, the number of European nation-states is increasing. Such nationalisms, however, can no longer simply be based on an unsatisfactory Wilsonian ideology and programme, due to the sharply diminished relevance of the 'nation-state' to the economic and geopolitical structure of the globe. Rather, in both halves of Europe we find aspiring nation-statelets going simultaneouly national and supranational. Even for the most insular of East European nations, access to the regionalization of West European markets through either full or associate member status of the European Community remains a goal.

REFERENCES

Agnew, J. 1983: An excess of 'national exceptionalism': towards a new political geography of American foreign policy. *Political Geography Quarterly*, 2, 151–66.

Agnew, J. and Corbridge, S. 1989: The new geopolitics: the dynamics of geopolitical disorder. In R. J. Johnston and P. J. Taylor (eds), *A World in Crisis?: geographical perspectives* (second edition), Oxford: Basil Blackwell, chapter 10.

Anderson, B. 1983: *Imagined Communities*. London: Verso.
Barnett, R. 1983: *The Alliance: America, Europe, Japan: makers of the post war world*. New York, Simon and Schuster.
Bell, D. 1961: *The End of Ideology*. New York: Macmillan.
Bell, D. 1973: *The Coming of Post-Industrial Society*. New York: Basic Books.
Bender, P. 1981: *Das Ende des Ideologischen Zeitalters*. Berlin: Severin and Sielder.
Bialer, S. 1986: *The Soviet Paradox: external expansion, internal decline*. London: I. B. Tauris.
Bunce, V. 1985: The empire strikes back: the evolution of the eastern bloc from a Soviet asset to a Soviet liability. *International Organization*, 39, 1–46.
Calleo, D. 1987: *Beyond American Hegemony: the future of the Western Alliance*. New York: Basic Books.
Callinicos, A. 1991: *The Revenge of History: Marxism and the East European revolutions*. Oxford: Polity Press.
Chase-Dunn, C. 1989: *Global Formation: structures of the world-economy*. Oxford: Basil Blackwell.
Chirot, D. 1985: The rise of the west. *American Sociological Review*, 50, 181–95.
Cohen, S. 1982: A new map of global geopolitical equilibrium: a developmental approach. *Political Geography Quarterly*, 1, 223–42.
Cox, R. 1987: *Production, Power and World Order: social forces in the making of history*. New York: Columbia University Press.
Dalby, S. 1990: American security discourse: the persistence of geopolitics. *Political Geography Quarterly*, 9, 171–88.
Dawisha, K. 1988: *Eastern Europe, Gorbachev and Reform: the great challenge*. Cambridge: Cambridge University Press.
Denitsch, B. 1990: *The End of the Cold War: European unity, socialism and shifts in global power*. London: Verso.
Deutscher, I. 1960: *The Great Contest: Russia and the west*. Oxford: Oxford University Press.
Folke, S. and Sayer, A. 1991: What's left to do?: two views from Europe. *Antipode*, 23, 240–8.
Frankel, S. 1983: *Beyond the State?: dominant theories and socialist strategies*. London: Macmillan.
Fukuyama, F. 1989: The end of history?. *The National Interest*, Summer.
Gellner, E. 1990: Perestroika in historical perspective. *Government and Opposition*, 25, 3–15.
Gill, S. and Law, D. 1988: *The Global Political Economy*. New York: Harvester.
Gowan, P. 1990: Western economic diplomacy and the new Eastern Europe. *New Left Review*, 182, 63–84.
Gray, C. 1977: *The Geopolitics of the Nuclear Era: heartlands, rimlands, and the technological revolution*. New York: Crane, Russack and Co.
Habermas, J. 1990: What does socialism mean today? the rectifying revolution and the need for new thinking on the left. *New Left Review*, 183, 3–22.
Hall, J. 1985: *Powers and Liberties: the causes and consequences of the rise of the West*. Harmondsworth: Penguin.
Halliday, F. 1983: *The Making of the Second Cold War*. London: Verso.
Halliday, F. 1990: The end of the Cold War. *New Left Review*, 180.
Hankiss, E. 1990: *East European Alternatives*. Oxford: Clarendon Press.

Harvey, D. 1989: *The Condition of Postmodernity*. Oxford: Basil Blackwell.

Hauner, M. 1990: *What is Asia to US? Russia's Asian heartland yesterday and today*. Boston, MA: Unwin Hyman.

Hirst, P. 1989: Endism. *London Review of Books*, 23.

Hobsbawm, E. 1990: *Nations and Nationalism since 1780: programme, myth, reality*. Cambridge: Cambridge University Press.

Johnston, R. J. and Taylor, P. J. (eds) 1989: *A World in Crisis?: geographical perspectives*. Oxford: Basil Blackwell.

Jones, R. 1990: *The Soviet Concept of 'Limited Sovereignty' from Lenin to Gorbachev: the Brezhnev doctrine*. London: Macmillan.

Jonsson, C. 1984: *Superpower: comparing American and Soviet foreign policy*. London: Pinter.

Kaldor, M. 1990: *The Imaginary War*. Oxford: Basil Blackwell.

Kaldor, M. 1991: After the Cold War. In M. Kaldor (ed.), *Europe from Below: an east–west dialogue*, London: Verso, 7–26.

Kennedy, P. 1989: *The Rise and Fall of the Great Powers*. London: Fontana.

Keohane, R. 1984: *After Hegemony*. Princeton, NJ: Princeton University Press.

Kobayashi, A. and Mackenzie, S. (eds) 1989: *Remaking Human Geography*. Boston, MA: Unwin Hyman.

Konrad, G. 1984: *Antipolitics: an essay*. San Diego: Horcourt.

Lang, O. (ed.) 1962: *Problems of Political Economy of Socialism*. New Delhi.

Lee, R. 1990: Making Europe: towards a geography of European integration. In M. Chisholm and D. Smith (eds), *Shared Space, Divided Space*, London: Unwin Hyman, 235–9.

Lewin, M. 1988: *The Gorbachev Phenomenon: an historical interpretation*. Berkley, CA: University of California Press.

Linklater, A. 1990: *Beyond Realism and Marxism: critical theory and international relations*. London: Macmillan.

Lovering, J. 1987: Militarism, capitalism, and the nation-state: towards a realist synthesis. *Environment and Planning D: Society and Space*, 5, 283–302.

Macmilian, B. (ed.) 1989: *Remodelling Geography*. Oxford: Basil Blackwell.

Malcolm, N. 1989: *Soviet Policy Perspectives on Western Europe*. London: Routledge.

Mann, M. 1988: *States, War and Capitalism*. Oxford: Basil Blackwell.

Materialy XXIV se"zda KPSS 1971: Moscow: Izadel'stvo Politicheskoi Literatury.

Parker, G. 1985: *Western Geopolitical Thought in the Twentieth Century*. London: Croom Helm.

Peet, R. and Thrift, N. J. (eds) 1989: *New Models in Geography* (2 vols). Boston, MA: Unwin Hyman.

Pinder, J. 1991: *The European Community and Eastern Europe*. London: Pinter.

Pugh, M. and Williams, P. (eds) 1990: *Superpower Politics: change in the United States and Soviet Union*. Manchester: Manchester University Press.

Rutland, P. 1989: Capitalism and socialism: how can they be compared?. In E. F. Paul et al. (eds), *Capitalism*, Oxford: Basil Blackwell, 197–227.

Sakwa, R. 1990: *Gorbachev and His Reforms 1985–1990*. London: Philip Allan.

Smith, G. 1989a: Privilege and place in Soviet society. In D. Gregory and R. Walford (eds), *Horizons in Human Geography*, London: Macmillan, 320–40.

Smith, G. 1989b: *Planned Development in the Socialist World*. Cambridge: Cambridge University Press.

Smith, G. (ed.) 1990: *The Nationalities Question in the Soviet Union*. London: Longman.

Smith, G. 1991: The state, nationalism and the nationalities question in the Soviet republics. In C. Merridale and C. Ward (eds), *Perestroika in Historical Perspective*, London: Edward Arnold.

Sokoloff, G. 1987: *The Economy of Detente: the Soviet Union and Western capital*. Berg: Leamington Spa.

Starr, R. 1985: *USSR Foreign Policies after Detente*. Stanford: Hoover Institution Press.

Steel, R. 1977: *Pax Americana*. Harmondsworth: Penguin.

Stent,? 1991: Gorbachev and Europe: an accelerated learning curve. In H. Balzer (ed.) *Five Years that Shook the World: Gorbachev's unfinished revolution*, Boulder: Westview.

Strange, S. 1987: The persistent myth of lost hegemony. *International Organization*, 41, 551–74.

Syzmanski, A. 1982: The socialist world-system. In C. Chase-Dunn (ed.) *Socialist States in the World System*, London: Sage Publications, 57–84.

Taylor, P. J. 1991a: The crisis of the movements: the enabling state as quisling. *Antipode*, 23, 214–28.

Taylor, P. J. 1991b: A theory and practice of regions: the case of Europe. *Environment and Planning D: Society and Space*, 9, 183–96.

Thompson, E. P. 1985: *The Heavy Dancers*. London: Merlin.

Thrift, N. 1989a: The geography of international economic disorder. In R. J. Johnston and P. J. Taylor (eds), *A World in Crisis?: geographical perspectives*, Oxford: Basil Blackwell, 16–78.

Thrift, N. 1989b: New times and spaces? the perils of transition models. *Environment and Planning D: Society and Space*, 7, 127–8.

Waever, O. et al. (eds) 1989: *European Polyphony: perspectives beyond east–west confrontation*. London: Macmillan.

Wallerstein, I. 1984: *The Politics of the World Economy: the states, the movements and the civilisations*. Cambridge: Cambridge University Press.

Worsley, P. 1984: *The Three Worlds: culture and world development*. London: Weidenfeld and Nicolson.

Z 1990: To the Stalin Mausoleum. *Daedalus*, 119.

Zaslavsky, V. 1982: *The Neo-Stalinist State: class, ethnicity, and consensus in Soviet society*. New York: Harvester Press.

5

Human Societies and Environmental Change: The Long View

I. G. Simmons

HOW LONG?

The title given me for this particular chapter is reminiscent of the phrase beloved of the French historian Ferdinand Braudel, '*la longue durée*'. For him, it was a long stretch of time in which personalities might be forgotten, as might individual events, in the greater swings and pulses of history. But that time generally started with a well-defined culture such as the Franks or the Romans; by contrast, I want to emphasize the time which starts when human societies began to alter their surroundings deliberately (or even accidentally if it was on a large scale), and that is less susceptible to accurate dating. (There is an even greater contrast between my standing and that of the founder/father figure of the *Annales* historians, and this essay is no attempt to emulate that kind of stardom.)[1]

But let us actually begin somewhat earlier. For example, we can situate ourselves in *cosmic time*. Human history cannot after all be divorced from the last 15 billion years of:

1 *creativeness*: in terms of the formation of galaxies, of DNA, and of the organic evolution of the complex forms found on this planet;
2 *conservatism*: the basic processes now evolved are difficult to envisage in any other form; for instance, the gravitational constants, and the laws of thermodynamics as they relate to the conservation of energy and the creation of entropy.

We have to be clear, of course, that there is no metaphysical reason why *Homo sapiens* should have an unlimited tenure in evolutionary terms. The rocks of the world are littered with the remains of species that have not made it beyond a certain time. Anthropocentrically, we tend to assume that we are in for a long period of hard work rather than for a short but noisy party.

Nested within cosmic time there is *humanity's time*. In this, the necessities of biology join with the evolution of symbolization to produce an animal which is both material and cultural and which possesses the power (both intellectual and technological) to alter the processes and forms of this planet, in the ways described in Andrew Goudie's chapter (chapter 6). A date like 10,000 years ago (k.y.a.) does not mark the beginning of humanity's time, for the history of humankind's putative effects on landscapes is long, even if subject to rather patchy evidence and a diversity of interpretation (Simmons, 1988). Nevertheless, as an arbitrary but not unreasonable starting date let us take the opening of the Holocene period (Roberts, 1989). This marks the beginning of that stretch of about 10,000 years since the ice began to wane in earnest from the temperate zones, during the whole of which human societies have possessed enough knowledge of nature and of themselves to be able to construct a technology which has allowed them to alter their environments cumulatively.

After examining briefly the types and quantities of change in that period, I want to turn to the meaning and interpretation of that change as we can see it today. In doing so, I shall go beyond the traditional materials of the geographer.

THE LAST 10,000 YEARS: THE COMPONENTS OF ECOLOGICAL CHANGE

Maps of the reconstructed land cover of the earth 10,000 years ago and today reveal some major differences. What must be explored is the extent to which those differences either are the outcome of 'natural' processes, such as climatic change and autochthonous ecosystem succession, or have been emplaced by human activity. With both in a state of flux and neither subject to systematic measurements of known accuracy, the certainties tend to vanish once the levels of either global-scale generalization or local monitoring have been passed, in any direction.

Climatic Change

Thanks to a great deal of work in a number of disciplines, we now have a clearer idea than ever of the course of climatic change during the last 10,000 years at a variety of spatial scales. What is clear is that:

1 the course of climatic change (in terms of, for example, temperature changes and biotic response thereto) shows considerable similarities to the previous interglacial periods of the Pleistocene;
2 this track is not, however, a simple bell-shaped curve of amelioration and retrogression (both words which need careful de-anthropocentricization) but has fluctuations of a smaller amplitude superimposed upon the great swings;
3 the order of fluctuation within the interglacials and the Holocene is not high in most measured values (for example, the difference in the receipt of solar radiation between 10 k.y.a. and now is about 8 per cent; in 10 k.y. the sea temperature was about 5 °C less than now; most climatic variables measured fall within the 1–10 per cent bracket);
4 these small changes in values can lead to larger-scale responses in cover: between the glacial maximum at 18 k.y.a. and now, the area of glaciers has shrunk to about 40 per cent of the greatest cover. Biomass changes for those zones which experienced allogenic succession between bare ground and other vegetation types in the period 10^7 k.y.a. are of the order of 520–600 per cent when measured as chlorophyll content per unit area.

In the years of technological spread and apparent dominance that have made up much of this century, the role of climate was progressively diminished as an agent of human affairs. Determinism became unpopular as a scientific theory, and its revival in the 1960s was in an ecological rather than a climatic form; today its profile is higher than ever within the context of the fact that human societies can affect it on more than a purely local scale.[2] Yet climate has continued to be a forcing function in the natural world during the Holocene, as well as exerting pressures on human activities. The spread of beech in Europe and North America during the latter part of the Holocene, for example, has been shown by Huntley and Webb (1989) to be closely related to climatic change in spite of all the land-use and land-cover changes going on as a result of human activity. Indeed, we might also argue that as the effect of the weather is now sometimes costed, its influence in some human domains is likely to be greater than ever, even if highly selectively.

Human-Forced Changes

The recent interest in the connectivities between land cover and use, economic activities, ocean function and atmospheric processes has led to an explosion of work on the ‘human impact on the environment’.

Systematic treatments sit alongside historical ones (often from the same publisher, with the same examples) and short popular accounts vie with huge, mega-tome collections of essays. As with the sire of them all, the W. L. Thomas volume of 1956, they are mostly empirical, and indeed not all escape the criticism of the Thomas volume – that it was in the end anecdotal, a problem addressed in Turner et al. (1990).

So the finding of a firmer basis from which to rate the human impact which is currently causing some concern in places high (academia) and low (government) is not all that simple. Data for the last 50 or so years are well developed, but before then the errors of estimation bulk large. Numbers are often presented on the basis of being 'pre-agricultural' without any regard for the different times of the development and spread of such an economy, or for the climate-driven changes occurring in the 'natural' vegetation. So some figures contrast with a 'natural' level, whereas others merely contrast with an earlier and apparently arbitrary datum.

1 Land cover Accepting these limitations, the difference between 'original' and present-day cover is in the range of 0.0 per cent for tundra (which cannot be literally true, but is an artifact of the resolution of the data-gathering and processing techniques used) to 19.1 per cent for grassland and 19.5 per cent for the non-tropical forests (Matthews, 1983). The average is 9.3 per cent, a meaningless figure. Notably, this data set does not include cities, which are estimated to cover about 3 per cent of the land surface. Global figures conceal large regional differences: in Burma between 1880 and 1980, arable land area increased by 273 per cent, compared with 214 per cent in the USA in the same period.

2 Process magnitudes There are a great many disparate data for 'before and after' magnitudes of process, with the times of before and after varying a great deal.

Those which are *land-based* exhibit, as might be expected, a wide range of values: human-caused erosion world-wide seems to be about $2\frac{1}{2}$ times the natural level, but the difference between forest and barren soil in Africa has been recorded at several times that level. The quantity of warm-blooded faunal species made extinct world-wide since 1600 hovers around 1 per cent. By contrast, the number of pairs of little ringed plover in Britain went up by 6,500 per cent between 1948 and 1960 due to the expansion of the gravel industry. Eggshell thickness in predatory birds in the UK went down 11.5 per cent between 1920 and 1952, but with disproportionate results on the numbers of sparrow hawks and peregrine falcons. A useful symbolic figure might be world terrestrial productivity appropriated by human use, which is calculated at 39 per cent.

Those relating to *fresh water* seem to run out higher than the land-based values, with phosphorus in English rivers going up 40,000 times between 1940 and 1968, and world water consumption going up 457 per cent since 1900. Lead in ice, which connects both land-based and atmospheric processes, increased by 20,000 per cent from 800 BC to AD 1950.

The human-derived changes in the *atmosphere* are well known and range from low numbers (like particulates at +18 per cent since the Industrial Revolution) to the very high, such as sulphur dioxide 1860–1985 at 4,625 per cent. Carbon emissions are also a modest 18 per cent above their 'natural' value. However, as with all such measurements in any environment, the actual quantity may belie the significance to the functioning of a process; similarly, where the changes are made may be as important as their magnitude. We cannot doubt that a diminution of the incidence of fog in Oxford by 42 per cent between the periods 1926–64 and 1965–80 has had a considerable impact on something or other.

Inferences

First, we cannot predict when, where and how much technology is going to produce how much ecological change: no global environmental impact assessment exists. Further, any changes now brought about by human societies take place in the context of an already altered set of ecosystems; it is rare for any society to be altering pristine nature. Further still, we know relatively little about the ways in which natural ecosystems respond to perturbations of all kinds (their resilience), let alone human-altered examples.

Nor is there any obvious relation between environmental changes brought about by humans and the anxieties that they cause, at any rate in information-rich societies. Thus we cannot predict how environmental changes will reverberate in human societies (see, for example, the discussion by de Freitas, 1991, of the 'greenhouse crises') and produce misgivings about: (1) the past – should we have done that?; (2) the present – should be doing this?; and (3) the future – what will happen if we do (do not do) that? These vibrations are the cultural context for our thinking about the human-environment relationship at present, and need further exploration.

In conventional terms, we talk of the perception of environment and of problems; less conventionally, I want now to discuss the more complex relationships among history, culture and meaning, suggesting that they are not separate and parallel evolutions, but intimately linked. A technology and a mythology, for example, are likely to go together. At the end, I want to see if the polarities which are so ingrained into the western world view are perhaps features which have outlived their evolutionary usefulness.

RESONANCE

This is probably the best term for the reverberations mentioned above: it is used by the sociologist Luhmann (1989) to emphasize that nature does not communicate directly with human species. But its processes (both the entirely biophysical and the human-affected) are the subject of human cognitions which do cause a set of reactions in human societies, and this is labelled *resonance*. The level of resonance seems to be a function of:

1. detachment from direct experience of the processes involved. This is directly related to the progress of urbanization and its associated phenomena, which interpose themselves between individuals and the biophysical world. This seems to be in tension with:
2. communication about the environment, especially in writing (of which this book is an example) and via telecommunications, particularly those of a visual nature, since our perceptions of 'nature' seem to be dominated by what we see. These communications seem to have the dual qualities of at once enhancing and distancing us from the natural world: consider the likely impacts of the 'virtual reality' illusion when it becomes common;
3. the intensity of information in these channels. Two outcomes, among others, suggest themselves. The first is that information overload occurs: who, for example, can watch all the environment-related programmes on European and North American television? The second is that a variety of meanings emerge and so the receiver is left with a confusion of messages: a programme about the delicate nature of an environment may well be followed by advertisements for some product which results in the perturbation of that environment.

In terms of conventional study, this resonance can be somewhat disaggregated, into:

1. the dynamics in space and time of natural, near-natural and cultural ecosystems;

2 the political ecologies which people erect within and alongside these ecosystems;
3 the cognitive mechanisms through which people perceive their relationship to the other two, and in which dissonance is possible: expressed attitudes (especially in writing) may not prove a reliable guide to actions.

Vaclav Smil (1987) makes the point that 'environmental problems' may not have changed a great deal between 1970 and 1982, but that our perception of them has. To him, the resonance here enfolds both human adaptation and the existence of the biosphere. So what we need to discuss at this point is whether environmental change of type (1) and cultural change of type (2) have, in the long view, proceeded interactively or on parallel tracks. To think in terms of wholes both spatially and temporally, we must search for a framework in which many ecological and cultural features can be reintegrated.

A RECONSTRUCTION

One attempt at an integration of these various features of the last 10,000 years is given as table 5.1 (pp. 108–9). The original author (Thompson, 1989a, 1989b) calls the wholes 'cultural ecologies' and has concentrated on the cultural features of columns 1–7; I have added other linkages with a more environmental focus in subsequent columns. The features of the table require little further explanation except perhaps to put dates to the main expressions of each cultural ecology: the Riverine phase is that typical of the ancient kingdoms of Egypt, Mesopotamia, Indus and northern China; the Mediterranean empires range from those of Greece and Rome to those of Spain and Portugal, and merge into the Atlantic, which at its height is the full flourishing of the Industrial Revolution but which clearly has its origins in earlier times. The table suggests that we are now moving rapidly into a dominance of the Pacific Rim (just as the Atlantic Rim held sway for so long) at the same time as a genuinely global ecology is emerging, a trend which needs no emphasis in this essay and is taken up in economic terms in the contribution by Peter Dicken (chapter 2).

The subtext of the table is that there are necessary connections between all these features and that they are not just coincidences; for example, that a particular type of political organization is inextricably linked with a science, a religion and a view of nature – what Foucault

(1972) called an epistème. There is no *a priori* reason why this should not be true, but it is difficult to show that it is so for the past, since the facts needed for such reconstructions are apt to be guided by the theoretical frame already erected. Nevertheless, there seem to be enough generally accepted relationships (which are analogous if not entirely coterminous with the idea of a world view or *Weltanschauung*) for the table to give us a guide to a longer view of culture and environment in which to situate ourselves today. In addition, we cannot overlook the possibility that all such constructions are pieces of text that resemble a soufflé: to be served quickly and then whipped away before it collapses.

But the rows of information also parallel a stimulating concept in intellectual history, namely the notions put forward, especially by Jantsch (1980) and Prigogine and Stengers (1985), of self-organizing systems. These take as metaphor the unpredictable wholes that result from evolving and interacting systems and can be invoked to explain, for example, the origin of life. The concept of chaos is clearly relevant here as well, since these systems too produce indeterminate outcomes from apparently determinate initial conditions; that is, stochastic behaviour occurs in a system governed by exact and unbreakable laws. These systems are creative yet irreversible (since they are bound up with entropy formation in the universe), and so fulfil the conditions of being at once creative and conservative. If this is so, then our present time seems to be one of the transition from the ways and concepts of Atlantic time to those of the Pacific and Global cultural ecologies. We are, as others have put it, at a 'turning point' or a 'hinge'. Thus our concepts of a changing world and of changing disciplines are within this context of a transition, a shift which appears to us to be very rapid, although this could be illusory since the shift to industrialism was, and is, probably equally fast. Be that as it may, this is a time of the separation of the curds and the whey, and the homogeneous milk may not be recreated unless by events of cosmic proportions.

INFERENCES

One of the most important features of a self-organizing system is the nature of the feed-back loops, both positive and negative. Presumably we in academia are part of these loops, since we are the channels for some of the resonances between the world of thought and that of the biophysical systems. Many such loops are doubtless important and it is conceivable

Table 5.1 Environmental relations of cultural ecologies

1	2	3	4	5	6
Coastal-montane	'Tribal'	Elders	Pictorial	Hierophantic	Pantheism
Riverine	City-state	Elite plus slavery	Script	Enumeration	Momentary possession
Mediterranean	Empire	Feudalism	Alphabet	Geometry	Surrender to authority
Atlantic	Industrial nation-state	Capitalism, socialism	Print	Dynamics	Commitment to belief
Pacific–global	Enantiomorphic	'Flexible accumulation'	Electronics	Transformations	Holistic group consciousness

Key
1 Type of cultural ecology
2 Polity
3 Sociopolitical organization
4 Communication mode
5 Dominant paradigm of science
6 Dominant type of religion

Source: Derived from the written form in Thompson, 1989a, 1989b.

that there are some we do not know about (Roszak, 1980; Sheldrake, 1988). Those that can form an agreed basis for discussion include:

1 the findings of the natural sciences. We need to note that these are always provisional even within a realist ontology. Given the dominance of the scientific paradigm in the Atlantic period, this loop cannot be undervalued. Neither should it be overrated, since it is one of the elements of the resonance and certainly not the only one.

7	8	9	10	11	12
Continuity with nature	Solar	Low, with spurt	Low, sporadic, not well known	Rarely survives in landscape	Extinction of climatically marginalized species
Nature as deity	Solar, wind, water	Low	High but concentrated	Often still foundation of present landscape	Salinification and waterlogging of soils
Nature as sacred or converted	Solar, wind, water	Low, beginning to rise	High and widespread	Mostly disappeared but still part of palimpsest, especially in low island environments	Deforestation and soil loss
Nature as resources	Add fossil fuels	High	Often intense, almost ubiquitous	Dominant landscape in many parts of high island environments spreading to low island environments	Changes in oceans and atmosphere at global scale
Nature as having intrinsic value	Add nuclear, revival of solar	High but slowing	Ubiquitous, dichotomy between intense manipulation and wilderness	As above but more closely scrutinized	Loss of genetic diversity; noise

Key (cont.)
7 Environmental attitudes
8 Energy sources
9 Population growth rates
10 Environmental manipulation: spread/concentration
11 Environmental manipulation: permanence
12 Environmental pathologies

2 the operation of what Charles Darwin called 'that imperious little word "ought"'; that is, the action of normative behaviour which goes by the name of ethics. We might note that in the west most well-known thinkers tackled the question of 'man in nature' until the nineteenth century, but that there seems to have been a gap until the post-1960 period, with the rise of environmental philosophy and environmental ethics (for example, Attfield, 1983; Brennan, 1990).

It is perhaps a British blindness (especially but not entirely in geography) that we have not been closely tied in to the development of arguments about such matters as the possibility of nature having intrinsic rather than instrumental value; that is, of all nature having a value in and of itself, rather than as a set of resources for human kind. Equally, the questions of justice in resource and environmental matters (for example, Durning, 1990) have only recently surfaced in our literature: we need to remember perhaps that Gaia's daughter was Themis, goddess of justice.

If there is one overriding inference that can be made from all these considerations it is that the 'social' is every bit as important as the 'natural' when the two are brought together in an inclusive cognitive framework, especially one not dominated entirely by the positivist epistemologies of the natural sciences.

CAUTIONS

Spinning such a web of words as this brings with it the obligation to keep asking whether the unsaid matters are as important as those which are made into marks on the paper. In this discussion there are notions which deserve more attention, such as:

1 the dangers of historicism. Layouts like table 5.1 look like an ineluctable progression of a Whiggish character. That the outcome might well be disaster has, however, been a consistent theme of the last 30 years. But it does surely teach us (following the analogy of the curds and the whey) that there is no such thing as equilibrium: like the universe itself, these self-organizing systems are always far from equilibrium. There is thus no returning to previous levels and, especially, there is no Golden Age to be sought, no Arcadian ideal to retreat to. We might quote T. S. Eliot's *The Dry Salvages* for our motto here: 'Not fare well / but fare forward, voyagers'.
2 how feedbacks and resonance work for us. One interpretation of our role as teachers and researchers is making nature transparent to us and ourselves transparent to others (and the other) in the hermeneutic sense used by Habermas (1972). So we have to make comprehensible the natural world as text (as delivered by science) but also as experience (as relayed by parts of the social sciences and by the humanities), all put together as an evolving set of processes. A motto for this might be the inscription on a thirteenth-century altarpiece in Toledo cathedral: *¡Caminantes! No hay caminar hay que camino.*

3 the *telos*. If we reject simple historicism then we need to acknowledge that in human affairs a purpose or *telos* is often espoused and forms part of the feedback mechanism. Without this, history does indeed become not perhaps bunk but a Procrustean bed. That *telos* has to conserve as well as create: here is a task to which we seem committed by our participation in discussions like this book. The difficulty for most of us is the discernment of the 'right' *telos*: perhaps, with Krishna in the *Bhagavadgita*, we have to let our reward be in the actions themselves, never in their fruits: a mode of thought especially problematic for westerners.[3] Alternatively, there are products of the imagination (see the very end of this chapter) that stimulate but do not restrict.

EXPLORATIONS

If most professionals now reject very largely the 'expedition to wild places' mode of carrying out geography, we cannot ignore the other explorations that are going on in other subjects which are concerned with the relations of humanity and nature. Some of these explorations will seem as strange to us as the habits of 'the natives' must have done to our forerunners. We ought perhaps to be readier to learn from the latter than we were from the former, a view reinforced in Peter Taylor's essay (chapter 9) in this book. Let us start by looking at some of the difficulties that stand in the way of any radical reconstruction of the humanity–nature nexus, viewed as a whole.

Outworn Divisions

We tend, in geography, to lament our ignorance of philosophy. By this we mean that we are unaware of all the richness and complexity of the western tradition and that we hope that somewhere there remains undiscovered an authority whose ideas will put us all to right (environmentalists have recently discovered Spinoza in this light, for example). But what could be possible is that all of these writers are unhelpful, in the sense that they are part of the post-Cartesian syndrome in which specialization of approach to the world has led to a variety of language games which have few words and grammars in common. Thus philosophy is part of the problem, not the resource for an answer. It is presumably a groping towards this realization that has made the post-structuralists want to remove the 'privileged' position of philosophy and make it just one of the lines of approach to a proposition, and not an *a priori* set of foundations (Lyotard, 1974).

All these divisions among professionals might lead to a consequence of importance for today's world. Any problem develops its own set of practitioners with their own paradigms, which legitimate their own existence and sequester the problem into a conforming and probably comfortable set of ideas and practices. This was at the heart of the notion of 'paradigm' in science as developed by Kuhn (1962), but is found in a more radical form in Michel Foucault's (1972) construction of *discourse*. It seems possible that the current approaches to human-environment problems could be construed as one of these discourses, with its own invisible college and ways of procedure both practical and intellectual. For example, the role of governments in supporting environmental science, whose results are then supposed to command social obedience, can be seen as a bounded discourse in these terms.

Fallacies

First is the fallacy that concepts are sufficient in themselves. There is always the problem of where to stop, raised *inter alia* by table 5.1. Are ideas ever disembedded? Conceptual frameworks of all kinds are always embedded in larger cultural frameworks. This leads to the usefulness of comparative work on how concepts and ideas function in their own frameworks and how they deal with contemporary reality, as well as seeing how they might cross-fertilize each other.

Second is the fallacy of the subject. Post-structuralists have taught us the notion of decentring the subject: there is no single 'I', but the numerous 'I's which take part in various linkages: as consumer, as teacher, as father and so on, with no *necessary* congruity between them all. In a less anti-foundationalist way, but still disturbing, there is the realization that the 'self' as we currently understand it is a concept perhaps only some 200 years old. The theories and the understanding of functioning of self-awareness are not well developed: what, for example, is the evolutionary significance of self-awareness, or of the 'meaning of meaning'? The fallacy, then, is the assumption that what we think we are *is* what we are: such a view is unlikely to be adequate, accurate or consistent unless a highly pragmatic philosophy is followed.

Together, these mean that the conventional 'cultural resources' are inadequate for the task of understanding what it means to be human (and therefore what some kind of separate entity called 'environment' might mean) in the world at present. We have to look for something which is beyond the old certainties and foundations of both east and west.

TRANSCENDING THE DUALITIES

In some of the discussion above, I have reproduced some of the inherent characteristics of today's discussions of the human–environment theme, namely that it polarizes around dichotomies. Examples of these are 'man' and nature, reason and emotion, analysis and intuition, technocentrism and ecocentrism, creators and created, knowledge and myth, subject and object, and of course good and evil. It is time to look more closely at some of these and evaluate their role in our cultural frameworks for environmental concern (Peters, 1989).

We have already assented explicitly to the idea of humans as both material, biophysical beings and bearers of culture. Imbalance between these two can lead us, for example, to suppose that only what is created by our minds and the tools developed out of our symbolizations is important. By contrast, the opposite leads to a kind of primitivism, a return to a pre-industrial society of a pastoralist, Arcadian nature. I fear that the cross-stitching on the shepherds' smocks was likely to have blood on it from the inevitable tuberculosis.

Going further into our own nature, we need to acknowledge that we are both emotional and rational creatures. The overemphasis on rationality is part of the divisions discussed above in connection with the limitations of philosophy as a guide to future actions. An ethics of rational principles has been all very well, and has no doubt kept many an academic off the streets, but it often fails to move us to action in the way that is characteristic of poetic metaphors, images and rituals. But to become a slave simply to emotion is equally unhelpful, since we can be persuaded that desires are desirable, wants are needs and luxuries are necessities.

Out of the idea that humans are, *mutatis mutandis*, creators as well as created by cosmic and humanized time there comes the notion that we can also control. This is clearly true up to a certain point. That point is the one where we seek to transcend the limits which all forms of evolution have provided: because of environmental concerns, those of biogeochemistry and of thermodynamics are especially in focus at present, but there are others. It would seem as if not all the idealistic reasoning in the universe can wish away these processes. Yet to be human is not to be apathetic: to take the view that any change is for the worse and that we are incapable of improvement is not consistent with our capabilities.

We need, perhaps, to remember that every global process which we have initiated began locally somewhere. An extension of these beliefs might lead us to a traditionalist fallacy, in which legitimate long-term cultural values are mistaken as true for all time and not as products of the evolution of a particular (and changing) cultural framework. The opposite is change for its own sake, producing the kind of perpetual cultural trauma that Alvin Toffler (1970) labelled 'future shock'.

The problem before us, then, is to affirm that all these traits are genuinely human and that it is the balance of them which is important. In this we seem to need to transcend the dualities which form the basis of so much of our approaches and actions. We are accustomed to thinking of these dualities as being in permanent conflict, with the need for us to be on the side of the conquering 'good', a strand much evident in the west since Zoroastrianism and cemented into place especially by the Protestant Christian cultures exported along with industrialism. The alternative seems to be a kind of dynamic balance rather than a world view which emphasizes conquest.

Granted, many eastern traditions have focused on a balance, as in the ying and yang elements within the greater whole of the Tao, but a colonization of the west by such ideas looks unlikely: if it did not happen in the 1960s it seems improbable now. New ways of cognition within the western tradition seem a better option.

WHERE NOW?

In the light of the above discussion, many commentators would urge humanity on a global scale to espouse an alternative to the metaphor of conquest; so what forms might it take? One is emerging in biophysical terms in the Gaia hypothesis: the earth and all its constituents as a single organic system. The view is exemplified in the visual image of 'Spaceship Earth' which is found on a thousand book covers. More radical commentators emphasize the necessity for a felt bonding between people and the earth: a far cry from the traditional ways of the geographer operating in the rationalist traditions of western scholarship.

This last consideration may look too far ahead for most academic geographers, but we can immediately acknowledge that there is a duality to which we are all party: that of acquiescing in approaches to the world primarily grounded in either the biophysical environment or human societies. We could move beyond that: not backwards to a kind of 'man–environment relations' geography which was designed to take up battle positions on the field of environmental determinism, but forward to a genuinely inclusive geography, which would be global in space and catholic in time.

In this, for a start, those approaching from the natural sciences would have to acknowledge that human societies do not simply respond in a 'rational' way to what scientists tell them is the case: their construction of nature is far more complex in its organization and dynamics, as is shown by O'Riordan (1987) in his work on the public perception of nuclear power development. On the other hand, those grounded in the social construction of nature will have to come to terms with the fact that there will be limits to this, imposed by cosmic and organic evolution.

To this end, geographers will need a *telos* like anybody else, and if it is not explicit then it will be a hidden agenda item anyway. So let me end by stating the kind of geography which we could be interested in promoting as well as studying: this formulation is that of Kenneth Boulding (1989), though it is not singular to his work.

> War would be eradicated, as slavery and duelling were virtually ended. The 7 per cent or so of the world economy now devoted to the war industry would be devoted to investment both in education and in the infrastructure and the capital required to increase human productivity and poverty would diminish sharply all over the world. The expectation of life would continue to increase, infant mortality and premature death would be sharply reduced, birth rates would fall correspondingly to the point where the human population stabilizes at a manageable level, whatever that may be. All the arts would be widely practiced and enjoyed, people would enjoy their lives, death would come at the right time and not be resented, although grief would continue as an important part of the human experience . . . institutions for preventing the pathologies of power would be strengthened. Even if this future has a probability of only 10 per cent, it is worth the effort.

A good road on which to be a traveller and on which, in the right company, to fare forward.

NOTES

1 I am not sure that I would actually wish to: Braudel's work at once delights (because of its easy facility over a wide range of time and knowledge) and annoys (because of its cavalier use of sources and many throwaway, take-this-for-granted type of statements)

me. Some of his later work strikes me as having discovered the existence of historical geographers and then trying to present their work as a new kind of history.

2 When I was an undergraduate in the late 1950s, what we heard about was microclimate in the Geiger sense of hedges and shelterbelts, and urban climates.

3 Witness ex-Prime Minister Margaret Thatcher's reinterpretation of the parable of the Good Samaritan.

REFERENCES

Attfield, R. 1983: *The Ethics of Environmental Concern*. Oxford: Basil Blackwell.

Boulding, K. E. 1989: *Three Faces of Power*. London and New Delhi: Sage Publications.

Brennan, A. 1990: *Thinking about Nature*. Athens, GA: University of Georgia Press.

de Freitas, C. R. 1991: The greenhouse crisis: myths and misconceptions. *Area*, 23, 11–18.

Durning, A. B. 1990: *Apartheid's Environmental Toll*. Worldwatch Paper no. 95. Washington, DC: Worldwatch.

Foucault, M. 1972: *The Archaeology of Knowledge* London: Tavistock Publications.

Habermas, J. 1972: *Knowledge and Human Interests*. London. Heinemann.

Huntley, B., and Webb, T. 1989: Migration: species' response to climatic variations caused by changes in the earth's orbit. *Journal of Biogeography*, 16, 5–19.

Jantsch, E. 1980: *The Self-Organizing Universe*. Oxford: Pergamon Press.

Kuhn, T. S. 1962: *The Structure of Scientific Revolutions*. Chicago: University of Chicago Press.

Luhmann, N. 1989: *Ecological Communication*. Cambridge: Polity Press.

Lyotard, J. F. 1984: *The Postmodern Condition*. Minneapolis: University of Minnesota Press.

Matthews, E. 1983: Global vegetation and land use: very high-resolution data bases for climate studies. *Journal of Climatology and Applied Meteorology*, 22, 474–87.

Peters, K. E. 1989: Humanity in nature: conserving yet creating. *Zygon*, 24, 469–85.

Prigogine. I., and Stengers, I. 1985: *Order out of Chaos*. London: Flamingo Books.

O'Riordan, T. 1987: The public and nuclear matters. In N. Geary (ed.), *Nuclear Technology International*, London: Sterling Publications, 257–63.

Roberts, N. 1989: *The Holocene: an environmental history*. Oxford: Basil Blackwell.

Roszak, T. 1980: *Person/Planet*. New York: Abacus Books.

Sheldrake, R. 1988: *Presence of the Past*. London: Collins.

Simmons, I. G. 1988: The earliest cultural landscapes of England and Wales. *Environmental Review* 12, 105–16.

Smil, V. 1987: A perspective on global environmental crises. *Futures*, 19, 240–53.

Thomas, W. L. (ed.) 1956: *Man's Role in Changing the Face of the Earth*. Chicago: University of Chicago Press.

Thompson, W. I. 1989a: Pacific shift. In J. B. Callicott and R. T. Ames (eds), *Nature in Asian Traditions of Thought*, Albany, NY: SUNY Press, 25–36.

Thompson, W. I. 1989b: *Imaginary Landscape: Making worlds of myth and science*, New York: St Martin's Press.

Toffler, A. 1970: *Future Shock*. New York: Random House.

Turner, B. L. et al. (eds) 1990: *The Earth as Transformed by Human Action*, Cambridge: Cambridge University Press.

6

Land Transformation

A. S. Goudie

The transformation of the face of the Earth as a result of human actions – land transformation – has fascinated geographers for many decades (see, for example, Thomas, 1956; Turner, Clark et al. 1990) and the publication of Marsh's *Man and Nature* (1864) was a major landmark. Yet although some of the leading practitioners of the discipline, notably Humboldt and Sauer, have seen this as a central concern of geography, most histories of the subject have given it relatively sparse attention, and some analyses have relegated it to a footnote (see, for example, Wrigley, 1965). In this chapter I want to build upon Simmons's historical perspective on land transformation (chapter 5), to examine some of the ways in which current human activities are transforming the surface of the Earth, to consider some of the implications of future global changes occasioned by possible global warming, to see what geographers (especially physical geographers) are doing to improve our understanding of humanly induced environmental changes, and to suggest ways in which this contribution can be strengthened. In essence I concur with Stoddart (1987) that the importance of land transformation is such that it should become a central, rather than peripheral, focus for the discipline.

HUMAN IMPACT IN PREHISTORY

As Simmons has so exhaustively and elegantly pointed out (Simmons, 1989, and chapter 5 of this book), recent studies have demonstrated that our prehistoric forebears were less innocuous than was previously thought. In particular, the deliberate use of fire for over a million years has been of fundamental importance in transforming some of the world's major biomes, including temperate grasslands, evergreen scrub, boreal forests and savannas (see, for example, Pyne, 1982). Fire gave to Palaeolithic hunters and gatherers technological power of immense significance even when human numbers and population densities were slight. The extermination of mega-fauna by Palaeolithic predators created an extinction spasm of unprecedented rapidity (Martin and Klein, 1984).

The tempo of change accelerated in the Mesolithic and the Neolithic, and recent analyses of sediment cores from peat bogs, swallow holes, lakes and the sea floors have indicated that rates of sediment yield and accumulation increased as the new technologies of pastoralism and agriculture were introduced and new areas colonized (see, for example, Metcalfe et al., 1989). The early herders and farmers of Britain may have contributed to such phenomena as the spread of *Alnus* (alder), the decline of *Ulmus* (elm), and podsolization and peat bog development in the Holocene (Roberts, 1989). In the Bronze and Iron Ages still further environmental modification manifested itself, and there is a large body of information from valley fills in Britain (table 6.1) which suggests that floodplain alluviation was greatly accelerated as erosion from cultivated hillslopes brought large amounts of topsoil and colluvium into low-lying areas. In arid areas, the introduction of irrigation created accelerated sedimentation, waterlogging and salinization as long as four to five thousand years ago (Jacobsen and Adams, 1958). Even the primitive smelting of ores may have contributed to land transformation, for early furnaces required inputs of large quantities of suitable firewood and charcoal. So, for example, many major erosional scars in Swaziland, southern Africa, locally called *dongas*, seem to occur in close proximity to furnace sites.

CHANGING PRESSURES FOR LAND TRANSFORMATION

Notwithstanding the achievements of humans in modifying the landscape in prehistoric times, recent decades have seen a transformation in the power of humans caused by their increasing numbers, their increasing consumption, and the application of a whole suite of new technologies.

Table 6.2, based on the work of Kates, Turner and Clark (1991), attempts quantitative comparisons of the human impact on ten 'component indicators of the biosphere'. For each component they defined total net change clearly induced by humans to be 0 per cent for 10,000 years ago (BP) and 100 per cent for 1985. They then estimated dates

Table 6.1 Accelerated sedimentation in Britain in prehistoric and historical times

Location	Source	Evidence and date
Howgill Fells	Harvey et al. (1981)	Debris cone production following tenth-century AD introduction of sheep farming
Upper Thames Basin	Robinson and Lambrick (1984)	River alluviation in Late Bronze Age and early Iron Age
Lake District	Pennington (1981)	Accelerated lake sedimentation at 5000 BP as a result of Neolithic agriculture
Mid-Wales	Macklin and Lewin (1986)	Floodplain sedimentation (Capel Bangor unit) on Rheidol as a result of early Iron Age sedentary agriculture
Brecon Beacons	Jones, Benson – Evans and Chambers (1985)	Lake sedimentation increase after 5000 BP at Llangorse due to forest clearance
Weald	Burrin (1985)	Valley alluviation from Neolithic onwards until early Iron Age
Bowland Fells	Harvey and Renwick (1987)	Valley terraces at 5000–2000 BP (Bronze or Iron Age settlement) and after 1000 BP (Viking settlement)
Southern England	Bell (1982)	Fills in dry valleys in Bronze Age and Iron Age

Table 6.2 Selected forms of human-induced transformation of environmental components: chronologies of change

A Quartiles of change from 10000 BP to mid-1980s

Form of transformation	Dates of quartiles[a]		
	25%	50%	75%
Deforested area	1700	1850	1915
Terrestrial vertebrate diversity[b]	1790	1880	1910
Water withdrawals[c]	1925	1955	1975
Population size	1850	1950	1970
Carbon releases[d]	1815	1920	1960
Sulphur releases[e]	1940	1960	1970
Phosphorus releases[f]	1955	1975	1980
Nitrogen releases[d]	1970	1975	1980
Lead releases[d]	1920	1950	1965
Carbon tetrachloride production[d]	1950	1960	1970

B Percentage of change by the times of Marsh and of the Princeton symposium (Kates et al., 1991)

Form of transformation	% change	
	1860	1950
Deforested area	50	90
Terrestrial vertebrate diversity[b]	25–50	75–100
Water withdrawals[c]	15	40
Population size	30	50
Carbon releases[d]	30	65
Sulphur releases[e]	5	40
Phosphorus releases[f]	<1	20
Nitrogen releases[d]	<1	5
Lead releases[d]	5	50
Carbon tetrachloride production[d]	0	25

[a] Calculations assume a baseline or pristine biosphere about 10000 BP and 100 per cent change as of the mid-1980s. Percentages refer to the total of the later or 100 per cent figure.

[b] Number of verterbrate species that have become extinct through human action since 1600. Does not include possible waves of Pleistocene and earlier Holocene human-induced extinctions because of continuing controversy over their nature and magnitude.

[c] Total amount of water now withdrawn annually for human use.

[d] Total mass mobilized by human activity.

[e] Present human contributions to the sulphur budget.

[f] Amount of phosphorus mined as phosphate rock.

by which each component had reached successive quartiles (that is, 25, 50 and 75 per cent) of its 1985 total change. They believe that about half of the components have changed more in the single generation since 1950 than in the whole of human history before that date.

During the twentieth century the human population of the planet has exploded. It was not until the time of George Perkins Marsh that the world's population reached 1,000 million – a process that had taken about 3 million years. By 1989 it had reached a figure of over 5,200 million. The time required to add 1,000 million people to the Earth's population has been reduced to just twelve years.

Likewise, world industrial production has grown more than fifty-fold over the past century, with four-fifths of this growth occurring since 1950. Urbanization has been equally rapid. The world's urban population has increased ten-fold, from around 100 million in 1920 to 1 billion today.

In addition to this vital demographic factor, one also needs to appreciate the high importance of the technological factor, for in the twentieth century sources of energy have been harnessed as never before, enabling an increasing scale and rapidity of environmental change. World commercial energy consumption trebled between the mid-1950s and 1980.

The importance of the harnessing of energy can be clearly seen in the context of world agriculture. At the turn of the century, more or less throughout the world, farmers relied upon domestic animals to provide both draft power and fertilizer. They were largely self-sufficient in energy. However, in many areas the situation has now changed, and fossil fuels are being extensively used to carry out such tasks as the pumping (or, in many cases, the mining) of water, the propulsion of tractors and the manufacture of synthetic fertilizers (which in many cases are polluters and eutrophicators). The world's tractor fleet has quadrupled since 1950 and as much as two-thirds of the world's cropland may be ploughed and compacted by increasingly large tractors.

In 1850 the total area of world cropland was about 538 million hectares. By 1980 it was three times that figure (*World Resources, 1988–89* 18). However, in spite of this increase and all that it has entailed for habitat loss, soil erosion and hydrological modification, the increase in cropland has not generally kept pace with the increase in human population. This in turn implies that the intensity, and thus the environmental impact, of agriculture has been magnified over this period. This is brought out in table 6.3, which provides data on the changes in land use and population that have taken place in South Asia between

Table 6.3 Changes in land use and population in South Asia

	1880	1980	% change
Bangladesh			
Land use (km²)			
Arable	7738	9147	+18.2
Open forest	1477	875	−40.8
Closed forest	1135	900	−20.7
Population (millions)	24.9	86.9	+349
N. India			
Land use (km²)			
Arable	534470	814530	+52.4
Open forest	228080	186050	−18.4
Closed forest	306850	170810	−44.3
Population (millions)	132.7	403.1	+304
Pakistan			
Land use (km²)			
Arable	104390	196620	+88.4
Open forest	97560	43700	−55.2
Closed forest	43970	23610	−46.3
Population (millions)	21.1	86.7	+411

Source: *World Resources, 1988–89*, 1988: 268.

1880 and 1980. Whereas the human population in the area has increased by around three to four times, the extent of arable land has only increased by between 18.2 and 88.4 per cent. In the same period the forested area has decreased by around a quarter to a half.

THE GROGGY EARTH

Human activities have now caused wholesale landscape transformation at the local, regional and continental scales, (Goudie, 1990a). The multitude of impacts is as follows:

Plants:
Deforestation
Bush encroachment
Desertification
Savanna expansion
Maquis heathland
Prairie
Invasion and explosion
Forest decline
Reduction in genetic diversity

Soil:
Salinization
Acidification
Water erosion
Wind erosion
Lateritization
Organic loss
Compaction

Hydrology:
Groundwater decline
Lake desiccation
Flood generation
Eutrophication
Gully deterioration
Thermal pollution

Atmosphere:
Climate change
Fog/smog
Aerosol loadings
Urban microclimates
Ozone depletion

Animals:
Decline in numbers and range
Extinction
Reduction in diversity
Invasion and explosion
Dwarfing

Geomorphology:
Weathering by polluted air and water
Channelization
Cratering
Earthquake generation
Subsidence
Slope failures
Coast recession and accretion
Permafrost decay
Sedimentation
Gully development
Sand dune reactivation

Having cleared large areas of temperate forest in the past few centuries, farmers and lumberjacks are now removing forests from the humid tropics at rates of around 11 million hectares per year, exposing soils to intense and erosive rainfall events and on average increasing rates of sediment yield six-fold. The world's rivers are being dammed by around 800 major new structures each year, transforming downstream sediment loads, while huge reservoirs impounded behind 300-metre-high dams are generating seismic hazards and catastrophic slope failures. Some of the world's largest lakes, most notably the Aral Sea in the former USSR, are becoming desiccated because of irrigation use and inter-basin water transfers at a near continental scale. Fluids, both water and hydrocarbons, are being withdrawn from beneath cities and farmlands with concomitant subsidence that may amount to 8 to 9 metres. Recreational vehicles and trampling feet are damaging areas subject to burgeoning tourism. The development of tundra areas is disturbing the thermal equilibrium of permafrost, leading to the expansion of thermokarst (subsidence) phenomena. Coastlines are being 'protected' and 'reclaimed' by concrete-happy engineers, often without due care for the unexpected consequences that may ensue. Finally, globally we are pumping at least 500 million extra tonnes of dissolved material into rivers and oceans each year, and we are acidifying precipitation, to the extent that some of it has the pH of vinegar or stomach fluid, thereby modifying rates of mineral release and rock weathering.

The changes that I have outlined so far as resulting from land transformation are serious in themselves. Of these, soil erosion is probably one of the most serious issues facing the world's population today; it threatens the resource base of farmers. Indeed, I am confident that over the coming decades it poses a much greater threat to the human race and to world food supply than does the much-trumpeted global warming. It is strange that this phenomenon received so little attention in the UK Government's White Paper on the environment, *This Common Inheritance* (HMSO, 1990).

However, another important point that needs to be made is that these miscellaneous changes, which take place at the local, regional and continental scales, are in many cases rendering the Earth 'groggy' with respect to the future potential effects of climatic change at the global scale (table 6.4). The reduction in groundwater levels, the desiccation of inland seas, the subsidence of permafrost, the accelerated retreat of coastlines, the stresses that beset coral reefs, and the occurrence of coastal flooding are all examples of change that, while currently in progress, will be accentuated in a warmer world. They are already rendering sensitive areas and environments prone to the likely consequences of global climatic change. I say this not to belittle the likely consequences of global climatic change. The reverse is the case. Because global changes are likely to work on a groggy earth, their consequences are likely to be compounded.

Above all, as a result of the escalating trajectory of environmental transformation it is now possible to talk about *global* environmental change. There are two components to this (Turner, Kasperson, et al., 1990): systemic global change and cumulative global change. In the systemic meaning, 'global' refers to the spatial scale of operation and comprises such issues as global changes in climate brought about by atmospheric pollution. In the cumulative meaning, 'global' refers to the areal or substantive accumulation of localized change, and a change is seen to be 'global' if it either occurs on a worldwide scale or represents a significant fraction of the total environmental phenomenon or global resource. Both types of change are closely intertwined. For example, the burning of vegetation can lead to systemic change through such mechanisms as carbon dioxide release and albedo change, and to cumulative change through its impacts on soil erosion and biotic diversity.

Table 6.4 Potential global effects of changes currently in progress

Phenomenon	Current human abuse	Potential global warming abuse
Groundwater reduction in High Plains	Overpumping by centre pivot	Increased moisture deficit
Desiccation of Aral Sea and associated dust storms	Inter-basin water transfers	Increased moisture deficit
Permafrost subsidence	Vegetation and soil removal, urban heating, etc.	Warming
Coastal retreat	Sediment starvation by dam construction and coastal engineering structures	Sea-level rise
Coral reef stress	Pollution, siltation, mining	Overheating, more hurricanes, fast sea level rise
Coastal flooding	Groundwater and hydrocarbon mining	Sea-level rise and more frequent storms

GLOBAL CHANGE

Although there are those who have expressed doubts (most notably Idso, 1989), in the last decade it has become apparent to workers in all disciplines that humans may now be capable, for the first time in their three-million-year history, of altering the Earth at the global scale because of the emissions of gaseous materials into the atmosphere (Houghton, Jenkins and Ephraums, 1990). The social and economic consequences of such changes have been well discussed by Parry (1990, and chapter 7 in this book), but I would like to stress the ramifying geomorphological consequences should global warming occur. These are as follows:

1 *Vegetation controls:*
   (a) major changes in latitudinal extent of biomes;
   (b) reduction in boreal forest, increase in grassland, etc.;

   (c) major changes in altitudinal distribution of vegetation types (*c*.500 metres for 3 °C);
   (d) growth enhancement by CO_2 fertilization.
2 *Hydrological:*
   (a) increased evapotranspirational loss;
   (b) increased percentage of precipitation as rainfall at expense of winter snowfall;
   (c) increased precipitation as snowfall in very high latitudes;
   (d) increased risk of cyclones (greater spread, frequency and intensity);
   (e) changes in state of peatbogs and wetlands;
   (f) less vegetational use of water because of increased CO_2 effect on stomatal closure.
3 *Coastal:*
   (a) inundation of low-lying areas;
   (b) accelerated coast recession;
   (c) changes in rate of reef growth;
   (d) spread of mangrove swamp.
4 *Cryospheric:*
   (a) permafrost decay, thermokarst, increased thickness of active layer, instability of slopes, river banks and shorelines;
   (b) glacier melting;
   (c) sea ice melting.
5 *Aeolian:*
   (a) increased dust-storm activity and dune movement in areas of moisture deficit.

Vegetation, which is such a major control on the operation of geomorphological processes (Thornes, 1990), will be severely modified if global warming takes place to any appreciable degree. Changes will occur altitudinally (by *c*.500 metres for a 3 °C temperature rise) and latitudinally (see, for example, Emmanuel, Shugart and Stevenson, 1985). Moreover, elevated CO_2 levels may cause growth enhancement of trees at high altitudes (La Marche et al., 1984), and reduced water use in arid areas. Because of the exceptionally large degree of warming that it is anticipated will take place in high latitudes, the boreal forests of the northern hemisphere may be especially severely modified (Sargent, 1988). Further, if they can cope with rising sea levels, mangrove swamps (the distribution of which is strongly temperature dependent) may be expected to expand.

Global warming and associated changes in precipitation regimes and distributional patterns will have a series of hydrological consequences that may have geomorphological significance. For example, some workers have suggested that because tropical cyclones are highly temperature dependant, they may increase in frequency, intensity and extent in a warmer world (see, for example, Emanuel, 1987). Any increase in cyclone frequency and intensity would have numerous geomorphological and human consequences, including accentuated river flooding and coastal surges, severe coast erosion, accelerated land erosion and siltation, and the killing of corals because of siltation and freshwater effects (De Sylva, 1986). However, the Intergovernmental Panel on Climate Change (Houghton, Jenkins and Ephraums, 1990) is equivocal as to whether global warming will in reality cause a change in cyclone behaviour. They find little evidence of any increasing trend in the twentieth century in spite of the warming that has occurred so far. On the other hand, there is some evidence that cyclone activity declined during the cold years of the Little Ice Age (Spencer and Douglas, 1985).

Substantial hydrological changes may occur in other major environmental zones, and significant runoff changes may be anticipated for semi-arid environments, such as the south-west of the USA. For example, Revelle and Waggoner (1983) suggest that in the event of there being a 2 °C rise in temperature and a 10 per cent reduction in precipitation, water supplies would be diminished by 76 per cent in the Rio Grande region and by 40 per cent in the Upper Colorado. They demonstrate that a 2 °C rise in temperature would be most serious for water supplies and runoff in those regions where the mean annual precipitation is less than about 400 mm.

In areas affected by snowfall today, the changes brought about in a warmer world may be especially marked. There will be a tendency for a substantial decrease to occur in the proportion of winter precipitation that falls as snow. Furthermore, there will be an earlier and shorter spring snowmelt. The first of these two effects will cause greater winter rainfall and hence winter runoff, since less overall precipitation will enter snowpacks to be held over until spring snowmelt. The second effect will intensify spring runoff, leading to additional adverse consequences for both summer runoff levels (Gleick, 1986) and spring and summer soil moisture levels (Manabe and Wetherald, 1986).

On the other hand, in high latitude tundra environments warmer winters may cause more snow to fall, thereby creating increased runoff levels in the summer months (Barry, 1985). Indeed, Budyko (1974:242) predicts that because of increased precipitation (perhaps by as much as 500–600 mm per annum in the tundra zone), runoff in the former USSR north of 58–60 °N will increase by a factor of between two and three.

Changes will also take place in the cryosphere. Permafrost will melt in a warmer world, creating expanded thermokarst conditions. For North America the southern limit of permafrost will be displaced northward by 100–250 km for every 1 °C rise in temperature (Barry, 1985). However, the speed with which permafrost will degrade is a matter of uncertainty. It is probably a relatively slow process, so that permafrost will continue to exist in current areas of *continuous* permafrost. In areas of less continuous permafrost the rate will vary according to material conductivity. Also important is the nature of snow cover and of the vegetation layer. Their changing state in a warmer world may modify the direct consequences of warmer surface temperatures (Boer, Koster and Lundberg, 1990). Where the permafrost is ice rich or contains massive ground ice, subsidence and settling due to thawing will occur, with severe consequences for engineering structures. Furthermore, thaw settlement will induce a thermokarst relief, which can alter drainage patterns and change the course of streams. Thus some areas may become swampy, inhibiting human activities like farming and mining. With permafrost degradation, slope stability would also decrease and the active zone become thicker. Erosion of river banks and reservoir shorelines may increase, causing an augmentation in sediment loads. Coastal retreat will also gather momentum as permafrost degrades in coastal lowlands, and large areas may be inundated as elevations are lowered to below sea level due to thaw settlement, especially if combined with a rise in sea level.

Snowlines will rise too, and calculations for New Zealand suggest that if surface temperatures rise by 3.6–6.3 °C, snowlines will rise by at least 300 metres, but perhaps by 500 metres, vertically. Small glaciers will disappear from many mountain ranges, and New Zealand will probably see the loss of about 1,000 out of its total of over 3,000 (Chinn, 1988).

With a 3 °C increase in temperature in Austria, little more than half of today's glacier surface will remain ice-covered. Especially numerous and large debris flows are formed in periglacial areas, where the marginal parts of small glaciers and Alpine permafrost with abundant masses of fresh, non-consolidated and vegetation-free debris are predisposed as starting zones. With a modest rise in temperature, glaciers will shrink and glacierets will disappear, thereby uncovering moraine-filled couloirs on steep slopes. With the degradation of Alpine permafrost, the involved debris will drastically alter its hydrological and geotechnical properties.

Considerable controversy surrounds the question of what will happen to ice near the Poles. It is not clear, for example, what degree of temperature increase is required to melt Arctic sea ice. At one extreme Budyko (1974:223) thinks a temperature rise of 2–3 °C will suffice to remove many-year ice, whereas at the other Bentley (1984) believes a 10 °C increase would be required to remove Arctic sea summer ice. The values of Flohn (1982) and Parkinson and Kellogg (1979) lie somewhere in between. Uncertainties also surround the future of the West Antarctic ice cap, with views ranging from the catastrophic (such as Mercer, 1978) to the more sanguine (Robin, 1986).

The final aspect of anticipated geomorphological change in a warmer world is that associated with sea-level rise, caused by the melting of land ice and the thermal expansion of the upper layers of the ocean. The degree of rise that may take place will not necessarily be uniform across the globe, because of geoidal and tidal effects (Goemans, 1986; Clark and Primus, 1987), and storage in large reservoirs may reduce the rate of rise (Newman and Fairbridge, 1986). Moreover, as we have seen, considerable uncertainty also results from the problem of predicting the behaviour of Antarctic ice. Will, at one extreme, the West Antarctic ice sheet melt precipitously, or will warmer conditions in polar regions permit greater accumulation of ice as a result of augmented snowfall? The Intergovernmental Panel for Climate Change takes a relatively conservative view, and suggests that a rise of more than one metre over the next century is unlikely (Houghton, Jenkins and Ephraums, 1990).

However, even a rise of only a metre could have a wide range of consequences. This has been graphically described by the Dutch Ministry of the Environment (Hekstra, 1989)

> The length of coastline in the world is between 500,000 and one million kilometres. A rise in sea level of approximately one metre could potentially affect all land up to the 5 metre contour line, if maximum storm surges and the effects of salt water intrusion along river mouths are taken into account. The area potentially affected is thus of the order of five million km^2 – about three per cent of the land area of the globe, but one third of the total area of cropland in the world. Much of the threatened land is densely populated and includes many large cities: indeed as many as one billion people may be at risk.

The general effects of accelerated coastal submergence have been summarized by Bird (1986), and include coast recession, beach narrowing, platform submergence, marsh retreat, inlet enlargement, barrier transgression and lagoon expansion. The effects of global sea-level rise will be compounded in those areas that suffer from local subsidence as a result of local tectonic movements, isostatic adjustments and fluid abstraction. For example, the Nile Delta is threatened not only by sea-level rise but also by tectonic subsidence and reduced sediment supply consequent upon the damming of the Nile at Aswan and elsewhere, and a strip of coast 30 km wide may be lost by 2100 (Stanley, 1988).

Coral reefs have often been seen as environments that may be adversely affected either by increased temperatures, which can cause 'bleaching' of corals, or by sea-level rise. However, at the lower rates anticipated by the Intergovernmental Panel on Climatic Change, corals will probably be more than able to keep up with sea-level rise (unless their health has been damaged by human actions). Some atolls, because of their Holocene sea-level history, are sufficiently elevated that they may well withstand a couple of metres of sea-level rise (Spencer, 1991).

The relationship between sea-level rise and beach erosion is a matter of considerable importance and some uncertainty. In general, following the so-called Bruun Rule, such erosion is to be expected, but the precise amounts are difficult to quantify. However, on sandy beaches, exposed to ocean waves, the coastline may erode by the order of 100 metres if there is a 1 metre rise in sea level (Committee on Engineering Implications of Changes in Relative Mean Sea Level, 1987).

GEOGRAPHERS AND LAND TRANSFORMATION

It is evident from what has been said so far in this chapter, and in the contribution of Parry (chapter 7), that land transformation and environmental change have become of central importance if one is to understand or predict many aspects of either the environment or the global economy. However, it is not yet evident that geographers are making the contribution in this area that the nature and history of the discipline might cause one to expect. In table 6.5 I present some data from four major journals for the period from 1987 to 1990 inclusive: these show the percentage of papers published there that deal with environmental change and land transformation. In the *Geographical Journal* it is *c.*13 per cent, in the *Annals of the Association of American Geographers* it is 13.6 per cent, in the *Transactions, Institute of British Geographers* (NS) it is 5.5 per cent and in *Area* the figure is less than 2 per cent. I am also not sure that we are playing our role in international fora dealing with such matters. For example, geographers only represent about 5.6 per cent of those who have played a role in the planning of the International Geosphere-Biosphere programme, and a third of those come from just two countries (the erstwhile USSR and Poland).

Table 6.5 Articles dealing with environmental change and land transformation in four major journals, 1987–90

Journal	Volume	Year	Number of articles	Number dealing with change
The Geographical Journal				
	153	1987	24	3
	154	1988	22	2
	155	1989	22	3
	156	1990	24	4
Total			92	12 (13.04%)
Annals of the Association of American Geographers				
	77	1987	35	7
	78	1988	32	3
	79	1989	32	3
	80	1990	26	4
Total			125	17 (13.60%)
Transactions, Institute of British Geographers (NS)				
	12	1987	34	1
	13	1988	32	2
	14	1989	30	0
	15	1990	31	4
Total			127	7 (5.51%)
Area				
	19	1987	28	2
	20	1988	25	0
	21	1988	25	0
	22	1990	28	0
Total			106	2 (1.9%)

Likewise, a recent Economic and Social Research Council directory on global environmental research centres in the UK, based on a questionnaire sent out to UK departments in institutions of higher education, demonstrated that there was great activity in this area but that most of it was not being undertaken by geographers. Of the environmental change centres that were being or had been established, only three were dominated by geographers (East Anglia, Oxford and the School of Oriental and African Studies: ESRC, 1990). Economists appeared to be the dominant disciplinary group conducting such work in most universities.

Another survey, conducted by Professor Jim Rose, and submitted to the Natural Environment Research Council, identified 68 respondents in UK geography departments who were working on global change (Rose, 1990). This is a relatively small proportion of the total membership of the geographical profession, and many of these respondents were undertaking rather traditional palaeo-studies. There was very little activity evident in the study of phenomena like soil erosion, desertification and acidification, and only one respondent admitted to undertaking modelling work in this area.

Why is there this limited involvement in the study of global environmental change by British geographers? Various reasons can be advanced. First, the number of biogeographers and climatologists in geography departments has been so slight that it has precluded an effective contribution to the study of atmospheric and biospheric change. Likewise, as a discipline we have suffered from fissiparism and fragmentation at a time when the most important issues have demanded endeavour at the interface between human and physical geography. Many geomorphologists have gone off to study their dreikanters and eskers, while some social geographers have gone off to study issues that are neither durable nor very serious. Many geographers have retreated from a concern with the environment *per se* and the real world *per se* and have rapidly taken up and then dropped ephemeral concerns. As Stoddart (1987:334) averred:

> Quite frankly I have little patience with so-called geographers who ignore these challenges. I cannot take seriously those who promote as topics worthy of research subjects like geographic influences in the Canadian cinema, or the distribution of fast-food outlets in Tel Aviv. Nor have I a great deal more time for what I can only call the chauvinist self-indulgence of our contemporary obsession with the minutiae of our own affluent and urbanized society – housing finance, voting patterns, government subsidies for this and that, and how to get the most from them. We cannot afford the luxury of putting so much energy into peripheral things. Fiddle if you will, but at least be aware that Rome is burning all the while.
>
> We need to claim the high ground back: to tackle the real problems: to take the broader view: to speak out across our subject boundaries on the great issues of the day.

WHERE NOW?

In spite of the negative tone of the last section, there are many examples of geographers who have made distinguished contributions to the study of land transformation and global environmental change. There is a need to build upon such best practice and to use our traditional skills. Some approaches to environmental change are:

1 *The use of the past:*
   (a) palaeoclimatic reconstructions;
   (b) palaeoecological reconstructions;
   (c) historical geography of transformation.
2 *The study of impacts:*
   (a) physical;
   (b) cultural.
3 *The mapping and monitoring of change:*
   (a) where is change taking place?;
   (b) how fast is change taking place?;
   (c) is change cyclical or linear?
4 *The recognition of vulnerable people and areas:*
   (a) thresholds;
   (b) combinations of circumstance.
5 *Modelling and predicting of the future.*
6 *The combination of human and physical approaches to come up with solutions.*

One important area in which we have skill is in historical reconstruction of the environment. One thinks here of the work that has been done to develop palaeoclimatic reconstructions against which general circulation models can be tested (such as COHMAP: see COHMAP members, 1988), the efforts that have been made to use palaeoecological reconstructions to identify the causes and onset of acidification (such as Battarbee, 1988), the analyses of past climatic statistics (such as Walsh, Hulme and Campbell, 1988), the studies that historical geographers have carried out of some of those great Darbyesque themes such as the clearing of the woodland (Williams, 1989) or the conversion of wetlands (Williams, 1990), and the historical and geomorphological studies that can be used to demonstrate the nature, timing and frequency of natural climatic changes such as the Little Ice Age (Grove, 1988).

Also notable is the work that is being undertaken by, among others, a large team which was formerly at the University of Birmingham and is now located in Oxford, the Atmospheric Impacts Research Group (Parry, 1990), on the social and economic impacts of global warming. Such predictive studies might have been more widespread had we not retreated as a discipline so totally from the example of Huntington.

The mapping and monitoring of change is another fundamental contribution that we can make, using remote sensing, sequential maps or instrumentation. Notable here is the work of NUTIS at Reading University on monitoring vegetation change in Africa through satellite imagery (Millington and Townshend, 1988), and the work undertaken to establish rates of rainforest recession (Grainger, 1991). We still do not know the answer to some basic questions such as how fast rainforest is being removed or regenerated, and to what extent deserts are expanding. Indeed, we find it difficult to define what is meant by 'deforestation' and 'desertification'. The important study of Ives and Messerli (1989) on deforestation and denudation in the Himalayas clearly indicates that much discussion on the speed of tree removal and its consequential changes (landsliding, flooding in the Ganga lowlands, siltation of deltaic areas in Bengal) is based on sparse data and inaccurate assumptions. The historical data they provide on deforestation rates and landsliding events hit at perceived wisdom and demand changed attitudes to conservation and development.

We also need to employ our regional knowledge to identify susceptible and sensitive areas, critical thresholds and vulnerable people. We need to develop models that will help us in our task of prediction. Finally, we need to come up with solutions to problems. There is much, for example, that we can contribute to amelioration of desertification (see, for instance, Goudie, 1990b; Grainger, 1990) by means of certain technological responses based on expertise in applied geomorphology (Cooke and Doornkamp, 1990) or by the introduction of such techniques as social forestry.

Many in the outside world are expecting a lead from geographers. As Calder has written in his popular work *Spaceship Earth* (1991:12):

> Fresh modes of thought are needed, that escape from the narrow specialisms of recent decades. Geography is by long tradition the great integrator that links the village and the world, and addresses the inter-play between biophysical processes and the human factor. A new global geography is in the making, and the only question is how much of it the physicists will leave for geographers to do.

REFERENCES

Barry, R.G. 1985: The cryosphere and climate change. In M. C. MacCracken and F. M. Luther (eds), *Detecting the Climatic Effects of Increasing Carbon Dioxide*, Washington, DC: US Department of Energy, 111–48.

Battarbee, R. W. and 15 collaborators, 1988: *Lake Acidification in the United Kingdom 1800–1986*. London: Ensis.

Bell, M. L. 1982: The effect of land-use and climate on valley sedimentation. In A. F. Harding (ed.), *Climatic Change in Later Prehistory*, Edinburgh: Edinburgh University Press, 127–42.
Bentley, C. R. 1984: Some aspects of the cryosphere and its role in climatic change. *Geophysical Monograph*, 29, 207–20.
Bird, E. C. F., 1986: Potential effects of sea level rise on the coasts of Australia, Africa and Asia. In J. G. Titus (ed.), *Effects of Changes in Stratospheric Ozone and Global Climate*, Washington, DC: UNEP/USEPA, 83–98.
Boer, M. M., Koster, E. and Lundberg, H., 1990: Greenhouse impact in Fennoscandia – preliminary findings of a European workshop on the effects of climatic change. *Ambio*, 19, 2–10.
Budyko, M. I. 1974: *Climate and Life*. New York: Academic Press.
Burrin, P. J. 1985: Holocene alluviation in southeast England and some implications for palaeohydrological studies. *Earth Surface Processes and Landforms*, 10, 257–71.
Calder, N. 1991: *Spaceship Earth*. London: Viking.
Chinn, T. J. 1988: Glaciers and snowlines. In Ministry for the Environment (ed.), *Climate Change: the New Zealand response*, Wellington: Ministry for the Environment, 238–40.
Clark, J. A. and Primus, J. A. 1987: Sea level changes resulting from future retreat of ice sheets: an effect of CO_2 warming on climate. In M. J. Tooley and I. Shennan (eds), *Sea Level Changes*, Oxford: Basil Blackwell, 356–70.
COHMAP members 1988: Major climatic changes since 18000 yr BP: palaeoclimate observations and model simulations. *Science*, 241, 1043–52.
Committee on Engineering Implications of Changes in Relative Mean Sea Level, 1987: *Responding to Changes in Sea Level*. Washington, DC: National Academy Press.
Cooke, R. U. and Doornkamp, J. C. 1990: *Geomorphology in Environmental Management* (second edition). Oxford: Oxford University Press.
De Sylva, D. 1986: Increased storms and estuarine salinity and other ecological impacts of the greenhouse effect. In J. G. Titus (ed.), *Effects of Changes in Stratospheric Ozone and Global Climate. Vol 4: Sea Level Rise*, Washington, DC: UNEP/USEPA, 153–64.
Emanuel, K. E. 1987: The dependence of hurricane intensity on climate. *Nature*, 326, 483–5.
Emmanuel, W. R., Shugart, H. H. and Stevenson, M. P. 1985: Climatic change and the broad-scale distribution of terrestrial ecosystem complexes. *Climatic Change*, 7, 29–43.
ESRC (Economic and Social Research Council) 1990: Response to ESRC questionnaire on global environmental change research. Mimeo, 5 pages, July.
Flohn, H. 1982: Climate change and an ice-free Arctic Ocean. In W. C. Clark (ed.) *Carbon Dioxide Review 1982*. Oxford: Oxford University Press, 145–79.
Gleick, P. H. 1986: Regional water resources and global climatic change. In J. G. Titus (ed.), *Effects of Changes in Stratospheric Ozone and Global Climate. Vol. 3: Climatic Change*, Washington, DC: UNEP/USEPA, 217–49.
Goemans, T. 1986: The sea also rises: the ongoing dialogue of the Dutch with the sea. In J. G. Titus (ed.), *Effects of Changes in Stratospheric Ozone and Global Climate. Vol. 4: Sea Level Rise*, Washington, DC: UNEP/USEPA, 47–56.
Goudie, A. S. 1990a: *The Human Impact on the Natural Environment* (third edition). Oxford: Basil Blackwell.
Goudie, A. S. (ed.) 1990b: *Techniques for Desert Reclamation*. Chichester: Wiley.
Grainger, A. 1990: *The Threatening Desert: controlling desertification*. London: Earthscan.

Grainger, A. 1991: *The Tropical Rain Forests and Man*. New York: Columbia University Press.

Grove, J. M. 1988: *The Little Ice Age*. London: Methuen.

Harvey, A. M. and Renwick, W. H. 1987: Holocene alluvial fan and terrace formation in the Bowland Fells, Northwest England. *Earth Surface Processes and Landforms*, 12, 249–57.

Harvey, A. M., Oldfield, F., Baron, A. F. and Pearson, G. W. 1981: Dating of post-glacial landforms in the central Howgills. *Earth Surface Processes and Landforms*, 6, 401–12.

Hekstra, G. P. 1989: Consequences of a global rise in sea-level. *Ecologist*, 19, 4–13.

HMSO, 1990: *This Common Inheritance* (White Paper on the Environment). London: HMSO.

Houghton, J. T., Jenkins, G. J. and Ephraums, J. J. 1990: *Climate Change: the IPCC Scientific Assessment*. Cambridge: Cambridge University Press.

Idso, S. B. 1989: *Carbon Dioxide and Global Change: Earth in transition*. Tempe, AZ: IBR Press.

Ives, J. D. and Messerli, B. 1989: *The Himalayan Dilemma*. London: Routledge.

Jacobsen, T. and Adams, R. M. 1958: Salt and silt in ancient Mesopotamian agriculture. *Science*, 128, 1251–8.

Jones, R., Benson-Evans, K. and Chambers, F. M. 1985: Human influence upon sedimentation in Llangorse Lake, Wales. *Earth Surface Processes and Landforms*, 10, 227–35.

Kates, R. W., Turner, B. L., II and Clark, W. C. 1991: The great transformation. In B. L. Turner, W. C. Clark, R. W. Kates, J. F. Richards, J. T. Mathews and W. B. Meyer (eds), *The Earth as Transformed by Human Action: global and regional changes in the biosphere over the past 300 years*, Cambridge: Cambridge University Press, 1–23.

La Marche, V. C., Graybill, D. A., Fritts, H. C. and Rose, M. R. 1984: Increasing atmospheric carbon dioxide: tree ring evidence for growth enhancement in natural vegetation. *Science*, 225, 1019–21.

Macklin, M. G. and Lewin, J. 1986: Terraced fills of Pleistocene and Holocene age in the Rheidol Valley, Wales. *Journal of Quaternary Science*, 1, 21–34.

Manabe, S. and Wetherald, R. T. 1986: Reduction in summer soil wetness by an increase in atmospheric carbon dioxide. *Science*, 232, 626–8.

Marsh, G. P. 1864: *Man and Nature*. New York: Scribner.

Martin, P. S. and Klein, R. G. 1984: *Pleistocene Extinctions*. Tucson: University of Arizona Press.

Mercer, J. H. 1978: West Antarctic ice sheet and CO_2 greenhouse effect: a threat of disaster. *Nature*, 271, 321–5.

Metcalfe, S. E., Street-Perrott, F. A., Brown, R. B., Hales, P. E., Perrott, R. A. and Steininger, F. M. 1989: Late Holocene human impact on lake basins in central Mexico. *Geoarchaeology*, 4, 119–41.

Millington, A. and Townshend, J. R. G. 1988: *Biomass Assessment: woody biomass in the SADCC region*. London: Earthscan.

Newman, W. S. and Fairbridge, R. W. 1986: The management of sea-level rise. *Nature*, 320, 319–21.

Parkinson, C. L. and Kellogg, W. W. 1979: Arctic sea ice decay simulated for a CO_2-induced temperature rise. *Climatic Change*, 2, 149–62.

Parry, M. L. 1990: *Climate Change and World Agriculture*. London: Earthscan.

Pennington, W. 1981: Records of a lake's life in time: the sediments. *Hydrobiologia*, 79, 197–219.

Pyne, S. J. 1982: *Fire in America – a cultural history of wildland and rural fire*. Princeton, NJ: Princeton University Press.

Revelle, R. R. and Waggoner, P. E. 1983: Effect of a carbon dioxide-induced climatic change on water supplies in the western United States. In Carbon Dioxide Assessment Committee, *Changing Climate*, Washington, DC: National Academy Press, 419–32.

Roberts, N. 1989: *The Holocene: an environmental history*. Oxford: Basil Blackwell.

Robin, G. de Q. 1986: Changing the sea level. In B. Bolin et al. (eds), *The Greenhouse Effect, Climatic Change and Ecosystems*, Chichester: Wiley, 322–59.

Robinson, M. A. and Lambrick, G. H. 1984: Holocene alluviation and hydrology in the Upper Thames Basin. *Nature*, 308, 809–14.

Rose, J. 1990: Research on aspects of global environmental change in British Geography departments. Mimeo, 7 pages.

Sargent, N. E. 1988: Redistribution of the Canadian boreal forest under a warmed climate. *Climatological Bulletin*, 22, 23–34.

Simmons, I. 1989: *Changing the Face of the Earth*. Oxford: Basil Blackwell.

Spencer, T. 1991: Corals in the global greenhouse: response of a bio-geomorphic system. Paper presented at Institute of British Geographers' Conference, Sheffield, January 1991.

Spencer, T. and Douglas, I. 1985: The significance of environmental change: diversity, disturbance and tropical ecosystems. In I. Douglas and T. Spencer (eds), *Environmental Change and Tropical Geomorphology*, London: Allen & Unwin, 13–33.

Stanley, D. J. 1988: Subsidence in the northeastern Nile delta: rapid rates, possible causes and consequences. *Science*, 240, 497–500.

Stoddart, D. R. 1987: To claim the high ground: geography for the end of the century. *Transactions, Institute of British Geographers*, NS 12, 327–36.

Thomas, W. L. (ed.) 1956: *Man's Role in Changing the Face of the Earth*. Chicago: University of Chicago Press.

Thornes, J. B. (ed.) 1990: *Vegetation and Erosion: processes and environments*. Chichester: Wiley.

Turner, B. L., Clark, W. C., Kates, R. W., Richards, J. F., Mathews, J. T. and Meyer, W. B. (eds) 1990: *The Earth as Transformed by Human Action: global and regional changes in the biosphere over the past 300 years*. Cambridge: Cambridge University Press.

Turner, B. L., II Kasperson, R. E., Meyer, W. B., Dow, K. M., Golding, D., Kasperson, J. X., Mitchell, R. C. and Ratick, S. J. 1990b: Two types of global environmental change: definitions and spatial-scale issues in their human dimensions. *Global Environmental Change*, 1, 14–22.

Walsh, R. P. D., Hulme, M. and Campbell, M. D. 1988: Recent rainfall changes and their impact on hydrology and water supply in the semi-arid zone of the Sudan. *Geographical Journal*, 154, 181–98.

Williams, M. 1989: *Americans and their Forests*. Cambridge: Cambridge University Press.

Williams, M. 1990: *Wetlands*. Oxford: Basil Blackwell.

World Resources, 1988–89, 1988: New York: Basic Books.

Wrigley, E. A. 1965: Changes in the philosophy of Geography. In R. J. Chorley and P. Haggett (eds), *Frontiers in Geographical Teaching*, London: Methuen, 3–20.

7

Geographers and the Impact of Climate Change

Martin Parry

Of the several environmental changes which currently operate at the global or near-global scale there are three that imply potential changes of climate: increases in greenhouse gas concentrations in the atmosphere; acidification of rain and snow; and depletion of stratospheric ozone. Of these, greenhouse gas-induced warming has captured an increasing public awareness and geographers have played a significant role in estimating its physical, economic and social implications.

It has not, however, always been a smooth ride. Initial efforts by geographers to consider the possible connections between environmental and human change brought substantial criticism from colleagues in other disciplines. It has taken 50 years since the possiblist reaction to environmental determinism in the 1930s for geographers to have regained sufficient confidence to renew research in this area. This chapter reviews briefly some of the approaches that geographers have used to consider the implications of climate change, considers the contributions that they have made to redefining climate change in terms of its potential impacts, and concludes with some recommendations for areas of research in which geographers can make a prominent contribution.

CONCEPTUAL MODELS OF CLIMATE AND SOCIETY

In order to organize the potentially confusing array of relationships between climate and people into a useful analytic framework, a number of conceptual models have been developed by geographers. These have become more complex in recent years, thereby allowing an increasing variety of feedback processes to be described.

Impact Models

Impact models consider climate as causing impacts in a more or less linear, non-interactive fashion (Kates, 1985). The mechanisms are frequently treated as a single unordered set, the impacts being immediately contingent to the climatic event. This approach now tends to be criticized by some as simplistic and as having led geography down a deterministic path from which it has still barely recovered.

An illustration of this approach is the hypothesis of climatically induced settlement change as an explanation for the decline of the Mycenaean civilization between 1200 and 900 BC. Originally proposed by a radical archaeologist, it was taken up as an idea by geographers and climatologists (Carpenter, 1966; Bryson, Lamb and Donley, 1974). The abandonment of Mycenaean palaces and other settlements in the south Peloponnese has generally been attributed to invasion by the Dorians from the north-east. But, the argument goes, the Dorians may have been moving into an area that had already been depopulated by outward migration, induced by famine caused by recurring drought. Whatever the reason, there is archaeological evidence that the Mycenaean settlement shifted to the north-west Peloponnese. Any weakening of the polar high pressure at this time would have caused a more northerly track of the Atlantic depressions which generally bring spring and autumn rain to the Aegean. If this had occurred, then the only rainfall that would have continued to be reliable would have fallen on the western Peloponnese. The original Mycenaean settlement could have suffered persistent drought, and migration may have been the response.

However, there is little evidence to support this hypothesis of climatically induced migration: there is no evidence for the climatic changes *per se*, or for their possible effects on agricultural potential in the region. Nor is the alternative hypothesis of military overthrow refuted. The supposed connection between meteorological drought, changes in the resource base and shift of settlement has not been established.

This type of simple causal explanation characterized earlier studies of the relationship between climate and culture, studies now associated with the philosophy of environmental determinism which postulated that human actions were frequently ultimately determined by the natural environment. To illustrate, in 1907 Ellsworth Huntington argued for the existence of climatic 'pulsations', which periodically drove the nomadic peoples of central Asia to the fringes of the sedentary world in Europe and South West Asia (Huntington, 1907. Huntington's theory was an extension of that developed by the German geographer Brückner, 1890). He also suggested that Mayan migrations in central America and the desertions of Roman settlements in Syria were both in part a consequence of climatic shifts (Huntington, 1925). These ideas were subsequently rejected and it became fashionable in the 1940s and 1950s, at the height of the possiblist movement in geography, to ridicule Huntington's work.

However, a more reasonable conclusion today might be that Huntington's rather naive conclusions stemmed from insufficient data and inadequate analytical techniques rather than from unpromising hypotheses. The premise of a cyclical connection between climate and history still survives, but generally lacks sufficient rigour to be convincing. An example is the hypothesized relationship between a supposed 80-year temperature cycle, the rate of population increase and the degree of social disturbance (measured by a weighted index of historical chronologies: Takahashi, 1981). More recently, geographers have sought to construct hierarchies of climate–society impact models to trace the effects of a climatic event, such as a drought, as it cascades through physical and social systems. An example is the work by Warrick and Bowden (1981) tracing the impacts of drought occurrence in the US Great Plains.

Interactive Models

Geographers and others have introduced a greater realism into climate–society studies by considering adaptation and adjustment to climatic change. By adjustment is meant the short-term, conscious response to either a perceived impact or a risk of impact (for example, building a sea wall to prevent damage from a sea surge). By adaptation is meant the long-term, often intuitive response that embodies many sets of risks and opportunities (for example, systems of cultivation which, over the years, have matched agricultural activities to the local resource base: Burton et al., 1978). These two types of interaction allow consideration, first, of factors such as the vulnerability or resilience of different systems to impact, which in turn can affect how societies can (and do) respond to climate change (see, for example, Timmerman, 1986). Secondly, it introduces the issue of perception of potential impact, the risks or benefits involved and the opportunities to avoid or take advantage of them. Studies of perception have a long pedigree in geography and have recently been applied successfully to the climate change issue (Warrick and Riebsame, 1983).

An interesting example of this, taken from work by anthropologists rather than by geographers, is the role that climate may have played in the chronology of the Norse settlements that gained a foothold along the Greenland coast in AD 985 but were extinguished before AD 1500, probably between 1350 and 1450. In the thirteenth century the population, which then totalled about 6,000 in two settlements on the south and south western coast of Greenland, was subject to a synergistic interaction of stresses from hostile Eskimoes (Inuit), the decline of the European market for walrus ivory, and challenging climatic fluctuations, particularly during 1270–1300, 1320–60 and 1430–60:

> had these factors not coincided when they did, Norse Greenland very probably would have survived the fifteenth century and might well have endured to the present in some form. However, this sort of explanation treats human response to climatic stress as a minor and dependent variable . . . We must consider not only the nature of the external stresses that seem to have killed the Norse Greenland but also the reason for that society's selection of ultimately unsuccessful response to such stress. (McGovern, 1981)

The failure to adapt to changing environmental circumstances probably explains much of the Norse decline. They continued to emphasize stock raising rather than exploiting the rich seas around them, as the Inuit had successfully done.

There are two broad conclusions that may be drawn from this. First, most social and economic systems have adapted, often slowly over many years, so that they can absorb a certain range of exogenous perturbations (for example, short-term climatic variations) without a major disruption to the system. New adaptations are probably based on the experience of perturbations in the past and are likely to accommodate a certain perturbation range in the future. However, the entire range of future perturbation is unlikely to be accommodated because the cost of ensuring against very unlikely events is greater than the full (unmitigated) cost of the impact of those events. Current work by geographers on this subject has tended to build upon behavioural research by American geographers in the 1960s, such as Saarinen's (1966) study of farmers on the US Great Plains, whose strategies accommodated droughts up to a certain degree of severity and frequency but ignored droughts with a long return period.

A second conclusion is that while relatively frequent and less extreme anomalies can be absorbed without shock to the system, rare and extreme events may exceed its resilience, causing a major perturbation that can cascade into other subsets of the system (such as wider sectors of the economy). If this analysis is correct, then the impact of climate change comes not only from the average but also from the extreme event, and this has enabled geographers to redefine climatic change into forms that allow its potential social implications to be more readily analysed.

REDEFINING CLIMATE CHANGE

There are broadly three means by which geographers have sought to interpret climate changes as a change in resource issues affecting policy. Firstly, climate change can be seen as a change in the frequency of extreme events. This is based upon the hypothesis that impact comes not from the change in average resources available for society but from extreme events, and that society is more likely to respond to its perception of a change in frequency of these extremes (Warrick and Riebsame, 1983; Parry and Carter, 1984). To illustrate, few farmers (whether commercial or subsistence) plan activities on their expectation of the average return. They either gamble on good years or insure against bad ones. More risk-prone farmers tend to tune their activities to bad years, while commercial farmers tune them to good years. Moreover, the incidence of extremes can be markedly affected by relatively small change in mean climate. Thus, what was once a one-in-a-thousand-year event, such as the extreme dry and warm summer of 1976 in north-west Europe, could become a one-in-ten-year event with a mere 1 °C to 3 °C increase in annual temperature and a 10 per cent decrease in mean annual precipitation (Warrick, 1991).

Secondly, we can evaluate climate change in human terms as a change in level of risk; that is, in the probability of an adverse or beneficial event such as short-fall from critical level of output or excess above the expected yield. In agriculture, for example, we generally assume that farmers are entrepreneurs who base their expected return on a mix of good and bad years. This has important implications for the impact of climate change for two reasons: first, as outlined above, because linear changes in climate can cause quasi-exponential changes in the probability of crop failure or conversely bumper harvests; and secondly, because there are large parts of the world where these levels of risks are already critical. This is self-evident in the case of marginal farmers whose risks are already high. But amongst non-marginal farmers, the change of risk is likely to have a similar quasi-exponential effect on the risk of one type of farming being more profitable than another; that is, on its comparative advantage. Thus surprisingly large changes in economic activity (for example, in farming types) may be necessary to accommodate apparently small changes in average climate. An illustration of this can be drawn from the US Corn Belt, where maize is grown close to its maximum high-temperature tolerance limits. A run of 5 days with maximum temperatures above 35 °C can kill a maize crop; yet an increase of only 1.7 °C in mean annual temperature would probably triple the current likelihood of occurrence of such a heat wave (Mearns, Katz and Schneider, 1984). This magnitude of temperature change lies well within the realm of possibility over the coming 50 years under a 'business-as-usual' scenario of greenhouse gas warming.

Thirdly, since climate can reasonably be construed as a resource, climate change can be viewed as a change in resource opportunities or options, producing benefits or disadvantages that may require an adjustment to match altered resource levels. The mix of farming strategies that may be appropriate for the altered climate may be markedly different from the existing one (Parry, 1985).

A Conceptual Framework

Attempts have been made to develop a hierarchy of models to analyse the cascade of climate impacts through various elements of the biophysical, economic and social system. The pathways and linkages can be traced with broadly three sets of models – of climate changes, of climate impacts on physical potential, and of the downstream economic and social effects of these (figure 7.1). A number of geographers have been concerned with using outputs from atmospheric general circulation models as inputs to impact models in order to predict potential or actual economic responses to climate changes; and then to trace the downstream effect of these changes in potential using outputs from impact models as inputs to economic models (for example, regional input-output models). It is then possible to consider what policies might best mitigate certain impacts at specified points in the system (Kates, 1985; Parry, Canter and Konijn, 1988; Parry, 1990).

As in most issues of pollution abatement there are four broad types of possible response option to climate change:

1 Reduce the effluent output (for example, CO_2 and other radiatively active gases).

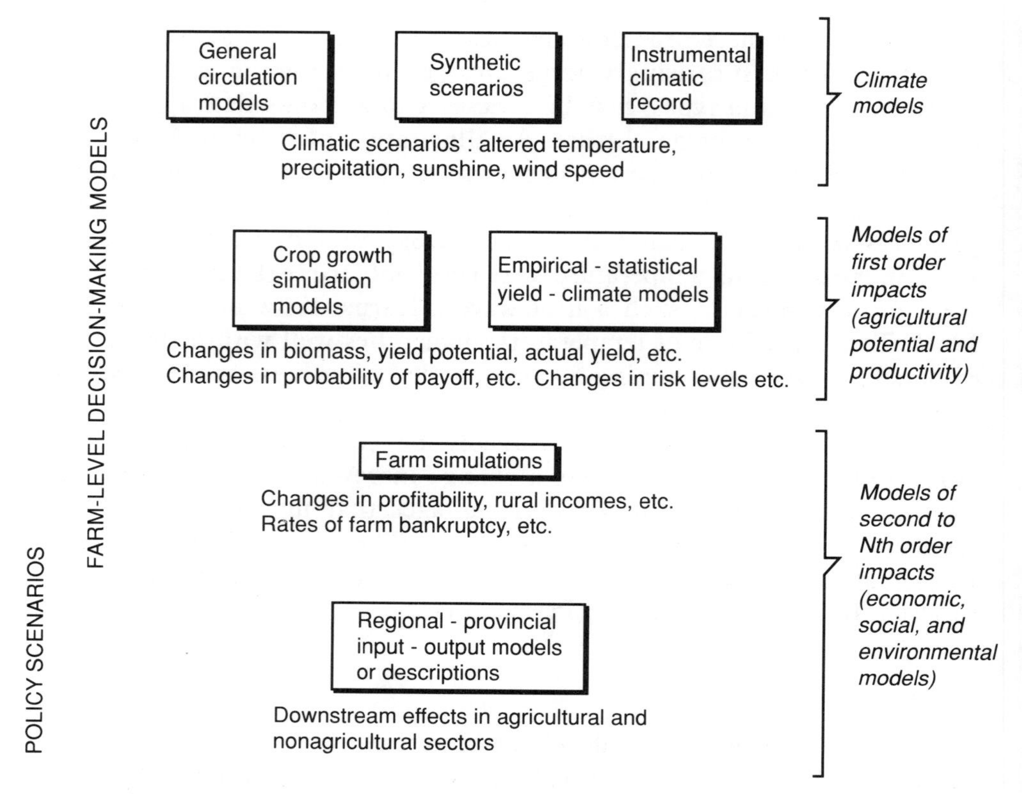

Figure 7.1 A hierarchy of models for the assessment of climate impacts and the evaluation of policy responses
(*Source*: Parry, 1986)

2 Reduce the polluting effects of the effluent (for example, remove CO_2 from the effluent or from the atmosphere.
3 Pay the price of making good the damage (for example, make countervailing modifications to climate, weather, or hydrology).
4 Suffer a reduced environmental utility (for example, adapt to increasing CO_2 and changing climate).

Within these broad lines of response there are, of course, many technological, economic and social devices to avoid damage, mitigate it or adapt to it. Most of the preventive or adaptive systems involve the transfer of resources from other sectors or other regions, and this presumes a certain degree of technological ability and economic integration. In general, the

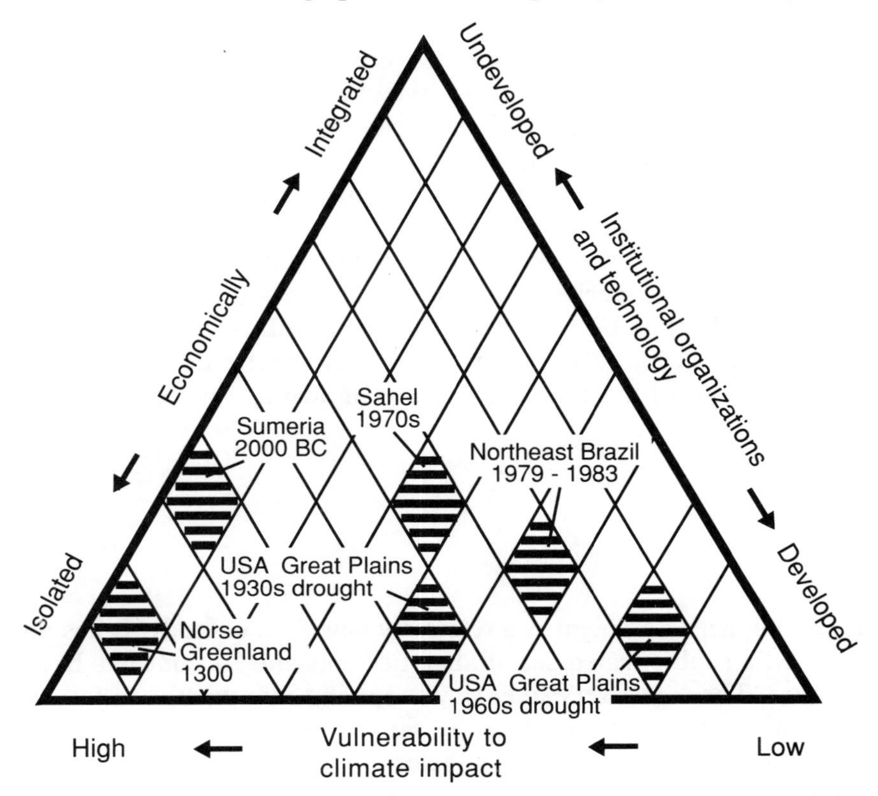

Figure 7.2 Technology, integration and the vulnerability of society to impact from climatic change
(*Source*: Parry, 1986)

greater the level of technological development and social organization the better are societies able to lessen the impact of minor climatic stresses (Bowden et al., 1981). At the same time, however, the degree of integration with other economies may affect the extent to which the impact is devolved or shared. To a partially closed economy, such as that of Norse Greenland in the fourteenth century, extreme climatic stress may cause a major shock to the system (figure 7.2). A similar, though temporary, impact was felt in the US Great Plains as a result of drought in the 1930s, when thousands of square miles on the dry margins of the Plains were effectively depopulated. By the 1960s, however, during which a substantial drought reoccurred, the impact was considerably less: increasing economic integration had served to devolve the impact of a climatic anomaly, while technology and social organization had made adaptation to it possible. But it is not yet clear how far such an increased resilience to drought would 'weather-proof' the economy of the US Great Plains against possible future and long-term changes of climate, particularly against pronounced increased frequencies of drought which, as we have seen, could result from relatively small changes in average conditions. There is a risk attached to this uncertainty. It may be wise, therefore, accepting that we cannot yet predict the future with a useful degree of accuracy, to consider the policies of response that would be required to a range of possible scenarios of future climate. When the picture becomes clear we should then be more ready to act accordingly. The alternative is to 'wait and see', but adaptive measures may require decades to research and develop, by which time the amount of adaptation required could be significantly greater.

CONCLUSIONS

Climate impact assessment is a relatively new field of study. It has not yet developed a sophisticated set of analytic tools and its concepts have still to mature. There is a very long way to go and many paths to follow, not all of which can be pursued at the same time. A selection of the most probably rewarding paths is given below.

First, we require more specific and user-oriented information regarding potential future climate change (its likelihood, magnitude, rate of onset, etc.). This information needs to be expressed in terms more readily applicable to the user (and often as derived parameters such as date of first and last frost, days of heat stress, etc.). Secondly, we require much more detailed study of the indirect effects of climate change (for example, via changes in soil chemistry, changes in the frequency of outbreak of agricultural pests and diseases, etc.). Thirdly, we need to specify with greater precision the interaction between climate and other resources in the primary sector, by modelling crop–climate, forest–climate and fisheries–climate relationships. It should then be possible to trace the downstream effects of these first-order impacts on the economy. Finally, we need to focus on the human side of the climate–human equation: constraints on choice, the role of decision-making, the enterprise level and the process of formulating strategies to cope with climate change. In all of these areas a premium will be placed on scientists who are able to integrate information from a number of traditionally separate disciplines. This is a task for which geography has been custom built.

REFERENCES

Bowden, M. H., Kates, R. W., Kay, P. A., Riebsame, and Weiner, D. 1981: The effect of climatic fluctuations on human populations: two hypotheses. In T. M. L. Wigley, M. J. Ingram and G. Farmer (eds), *Climate and History*. Cambridge: Cambridge University Press, chapter 21.

Brückner, E. 1890: Klimaschwungen Seit 1700, *Geographische Abhandlungen*, 4(2), 261–4.

Bryson, R. A., Lamb, H. H. and Donley, D. L. 1974: Drought and the decline of the Mycenae. *Antiquity*, 48, 46–50.

Burton, I., Kates, R. W. and White, G. F. 1978: *The Environment as Hazard*. New York: Oxford University Press.

Carpenter, R. 1966: *Discontinuity in Greek Civilization*. Cambridge: Cambridge University Press.

Huntington, E. 1907: *The Pulse of Asia*. New Haven, CT: Yale University Press.

Huntington, E. 1925: *Civilisation and Climate* (third edition). New Haven, CT: Yale University Press.

Kates, R. W. 1985: The interaction of climate and society. In R. W. Kates, J. Ausubel and M. Berberian (eds), *Climatic Impact Assessment: studies of the interaction of climate and society*, Wiley: New York, 3–36.

McGovern, T. H. 1981: In T. M. L. Wigley, M. H. Ingram and G. Farmer (eds), *Climate and History*, Cambridge: Cambridge University Press, chapter 17.

Mearns, L. O., Katz, R. W. and Schneider, S. H. 1984: Changes in the probabilities of extreme high temperature events with changes in mean global temperature. *Journal of Climatology and Applied Meteorology*, 23, 1601–13.

Parry, M. L. 1985: The impact of climatic variations on agricultural margins. In R. W. Kates, J. Ausubel and M. Berberian (eds), *Climatic Impact Assessment: studies of the interaction of climate and society*, Wiley: New York, 351–68.

Parry, M. L. 1986: Some implications of climate change for human development. In W. C. Clark and R. E. Munn (eds), *Sustainable Development of the Biosphere*, Cambridge: Cambridge University Press.

Parry, M. L. 1990: *Climate Change and World Agriculture*. London: Earthscan.

Parry, M. L. and Carter, T. R. 1984: *Assessing the Impact of Climate in Cold Regions*. Summary Report SR-84-1. International Institute for Applied Systems Analysis: Laxenburg.

Parry, M. L., Carter, T. R. and Konijn, N. T. (eds) 1988: *The Impact of Climatic Variations on Agriculture. Vols 1 and 2*. Kluwer: Dordrecht.

Saarinen, T. F. 1966: *Perception of the Drought Hazard on the Great Plains*. Research Paper No. 106. Chicago: Chicago University Geography Department.

Takahashi, K. 1981: Climatic change and social disturbance. *Geojournal*, 5, 165–70.

Timmerman, P. 1986: *Vulnerability, Resilience and the Collapse of Society*. Institute Monograph No. 1. Toronto: Institute of Environmental Studies, University of Toronto.

Warrick, R. A. 1991: Scenarios of climate change. In Department of the Environment, *Potential Effects of Climate Change in the United Kingdom* London: HMSO.

Warrick, R. A. and Bowden, M. 1981: Changing impacts of drought in the Great Plains. In M. Lawson and M. Baker (eds), *The Great Plains: perspectives and prospects*, Lincoln, NE: Center for Great Plain Studies.

Warrick, R. A. and Riebsame, W. E. 1983: Societal response to CO_2-induced climate change: opportunities for research. In R. S. Chen, E. Boulding and S. H. Schneider (eds), *Social Science Research and Climatic Change*, Reidel: Dordrecht, 44–60.

PART II

A Changing Discipline

8

Meet the Challenge: Make the Change

R. J. Johnston

This second introductory chapter assumes acceptance of the argument developed in the first part of the book: the world that geographers study is changing fast. Two linked questions follow: 'How should geographers respond to those changes, in terms of the orientation of their discipline – its concerns and content?'; and 'How should geographers seek to influence those changes?'. Possible answers are introduced here by a brief exploration of the discipline's nature, followed by a review of recent essays setting out certain individuals' views on geographers' roles in such a changing world. The three chapters that follow develop particular arguments in more detail.

GEOGRAPHY: ITS GOALS AND ITS METHODS

Much has been written about what geography is and is not, what it should be and what it should not.[1] Three sets of questions have been at the core of this literature: (1) 'What is geography's core material, and how does geography relate to other disciplines?'; (2) 'What philosophy should geographers adopt, and what methodologies are appropriate to that philosophy?'; and (3) 'What roles should intellectuals generally and geographers in particular seek to play within society – how should geography be presented in educational curricula, how should its material and methods be used, and how should geographers act?'. That substantial literature cannot be comprehensively reviewed here; the particular concern is with the response of geographers to current changes.

Applied Geography

The first part of this book has set out the contemporary context – economic, social and political restructuring and increasing environmental concern. Taylor (1985) has argued that the balance of 'pure' and 'applied' work undertaken within a discipline will reflect outside pressures. From geographers, who have 'never been strongly represented in the private sector' (p. 95) and whose main contributions have been made in 'the education part of the knowledge industry' (which is largely supported by public funds), changing economic conditions may call forth different responses:

> Outside pressures will be particularly acute in periods of economic recession when all public expenditure has to prove its worth. All disciplines will tend to emphasize their problem-solving capacity and we can expect applied geography to be in the ascendancy within our studies. In contrast in a period of expanding economies and social optimism outside pressures will diminish and academia can be expected to be under less external pressure. Geographers will be able to contemplate their discipline and feel much less guilty about this activity. (p. 100)

Applied geography should currently be in the ascendancy, therefore.

But what is applied geography? Pacione (1990:3) defines 'applied urban geography' as 'the application of geographical knowledge and skills to the resolution of urban social, economic and environmental problems'. Geographical investigations provide the basis for 'planning remedial action': they can also 'assist in the solution of an already perceived problem or can warn of an impending difficulty', through using a linked sequence of 'description, explanation, evaluation and prescription leading to implementation, followed by monitoring'.

This definition is closely tied to a particular vision of science, however, and thus of the nature of both problems and solutions. Elsewhere (for example, Johnston, 1986a, 1986b, 1990a, 1991a: see also Entrikin, 1991) I have followed Gregory (D. Gregory, 1978) and drawn on Habermas (1972) by identifying three types of scientific endeavour, each with its separate goals and thus applied value:

1 *Empiricist–positivist science and technical control*, in which prediction is equated with explanation – one can account for a phenomenon because one can state the circumstances in which it will and will not

occur (even if you do not know why: D. Gregory, 1980). With that ability, it should then be possible to avoid certain happenings, by eliminating their preconditions, and also to ensure that other events do occur, by engineering their precursors. Thus knowledge of this type allows for technical control – desired ends can be defined and attained. Some argue that this can be achieved with the physical world only, providing that valid predictive models are available; others contend that similar procedures can be applied both to individual humans and to the societies which they create – what is sometimes termed social engineering.

2 *Humanistic science and mutual understanding* has the goal of appreciating events involving humans by, for example, seeking to uncover the thoughts behind actions and to communicate the human condition. The only explanation sought is that provided, directly or indirectly, by the actors involved: analysts then communicate their understandings of both their own and others' experiences. Application of that understanding involves enhancing both self-awareness and mutual awareness; the researcher is not a technician promoting a certain solution but rather a provocateur who promotes thought and reflection, as a basis for action.

3 *Realist science and emancipation*: humanistic science provides appreciation of how people interpret the world as the basis for their actions, but that understanding may be flawed. People's perceptions of the world and of how it works may help them to survive, but if those perceptions are based on a false view they contain the potential either for people to be unable to cope in certain circumstances (the world does not respond to their actions as they anticipate) or for them to fail to tackle the real causes of any identified problems. Realist science attempts to eliminate such distorted views by identifying the underlying (structural) causes for events and actions. The nature of those causes can then be transmitted to people; by demonstrating to them how the world 'really works' they will then be emancipated – false interpretations are removed and people are given the ability to control their own futures.

These three different types of science call for very different types of geography within the 'knowledge industry' of higher education. The first requires an emphasis on the inculcation of technical and other skills which can be used in 'problem-solving' – the emphasis is strongly materialistic, stressing *technical training*. The other two call for *education*, defined here as a process of developing the individual and his or her appreciation of the world; this also requires training, in the qualitative methods of obtaining an appreciation of how people operate in their life-worlds and of transmitting that appreciation. (Interestingly, surveys of British graduates – including those with qualifications in geography: Johnston, 1990b, 1991b – indicate that whereas some retrospectively evaluated their university or polytechnic course in materialist terms with regard to the relevant labour market skills that it provided, others used criteria related to educational satisfaction.) Each type of science can have an applied component – to a greater or lesser extent, depending on the students' and teachers' interests. Whereas applied empiricist-positivist science involves attacking problems using an engineering approach, however, applied humanistic science is agnostic with regard to the future; it seeks to help people better create their own futures through greater self- and mutual understanding. Finally, applied realist science usually has a clear political goal, of helping people to create a world very different from that they are currently experiencing.

RESPONDING TO THE CHALLENGES

All three of these types of applied geographical science are being promoted at present, as ways in which geographers should respond to their changing world. The following exemplify the various arguments.

Empiricist-Positivist Modelling in a 'Post-welfarist World'

In a recent strong polemic, Bennett (1989) has both criticized much recent human geography for its implicit, if not explicit, political bias and argued for a new orientation more in tune with what he calls the 'culture of the times'. The discipline needs to open its theoretical perspectives to challenges so that it 'can contribute to dialogues on policy or education in a world of changing political, social and economic structures – a world which is identified, loosely, as moving towards a post-welfarist culture of the times' (p. 273). The approach to geography that Bennett finds wanting was developed during what he terms the 'welfarist culture of the times' which preceded the current rapid changes. Welfarism involved institutionalization of the responsibility of government; to ensure high and stable employment, to provide income support for the poor, to provide health care, education and housing, and to *plan* effective provision. Planning and rationality ... became by-words of the whole process and not least in economic management' (p. 277). He contends that this culture ensured that entitlements became institutionalized: a 'welfare state [developed] in which people have claims for welfare against society as a matter of *rights* or justice not merely as the outcome of altruism . . . needs thence become easily confused with wants and desires and the moral judgement as to the distinction of "want" and "need" breaks down' (p. 277).

The focus on relative deprivation within society which accompanied the growth of welfarism led to a concern with inequalities in many forms, including geographical. Further, given the perceived shift from expectations to rights then, since some differences exist because of geography, 'Geographical as well as other restrictions on choice, differences in lifestyles, differences in quality of life, access to certain goods and services can be identified. Because they can be identified, under a welfarist view, *morally* they should be overcome or at least ameliorated' (p. 279). The theories that geographers (basically human geographers) developed are termed 'social theory' by Bennett, who criticizes their proponents for allying geography with social science and thus 'losing its distinctive disciplinary focus' (p. 281: see below): some of these theories sought only to understand those differences whereas others promoted spatial engineering that would reduce, if not entirely eliminate, them.

Welfarism has failed, according to Bennett. Its logical conclusion is that every difference implies relative deprivation and thus an automatic entitlement to state action to remedy it, so that 'the only end point is total state intervention in everything' (p. 285). Under the post-welfarist system promoted in the United Kingdom in recent decades – and often termed 'Thatcherism' – however:

> capitalism has come to be seen as the means of creating and distributing the good things in life. Where social theory [see above] emphasizes the negative aspects of capitalism's capacity to create new wants and hence new 'relative deprivation', the emergent 'culture of the times' has been happier to see the market as both the creator and the provider of new wants. Rather than markets being seen as an inhumane and exploitative system, socialism and even corporatist social democracy have come to be associated with the odious and paternal treatment of individuals . . . Where the Thatcher era has heralded consumer choice and economic change, social theory and socialist politics has sought to defend the mode of production and to trap people in labour-intensive work practices and unattractive jobs vulnerable to technological change: the spirit of market freedom of individuals has heralded a consumer and

> service economy which has offered the release from the least attractive toils and labours, and has seemed to offer the potential to satisfy many of people's most avaricious dreams. (p. 286)

The glorification of the market which characterizes this post-welfarist view is coupled with a critique of the state's role as defined under welfarism:

> It is no longer credible intellectually or politically to argue for many previous forms of government management or intervention. Mass public housing, demand-inelastic systems of health care, education objectives which seek to equalize outcomes rather than potential, 'full employment' Keynesianism, regional policy, even the existence of local government, are each examples where policies have been developed over the past forty years, but where there have been significant and recurrent failures. (p. 287)

Thus the social theory which Bennett says human geographers have enthusiastically embraced has become redundant, he claims: 'even social democracy offers no easy solution'.

What, then, does Bennett see as geographers' role in this brave new world? 'Markets do fail' – even if they could provide for all our individual and quantitative needs, 'they certainly do not provide all our collective and qualitative needs' (p. 287). Thus:

> The two key questions for a post-welfarist society are: first how can support be improved by practical policies that can be demonstrated to work and are reasonably cost-effective; and second at what point does government action end. Should government be concerned merely with facilitative policies which allow human needs to be satisfied; or does it involve wider 'service' needs of income, food, shelter, health care, and so on. It is in the second area that it is clear that neither social theory nor pure market systems provide fully satisfactory guidance. (p. 287)

He calls for geographers to determine what is and is not possible through collective action, so that 'The grand objective of the discipline should be to contribute to the debates around these issues. But it must be a contribution to practice' (p. 288). Bennett sets out no research agenda, however, claiming lack of space and 'nor would I yet be that presumptuous' (p. 288). (Nevertheless, in his later edited book *Decentralization, Local Governments and Markets* – Bennett, 1990a – which has the subtitle *towards a post-welfare agenda* – he stresses 'the changes which are developing between *markets* and between *levels of government*' – Bennett, 1990b:25.) He does, however, argue that an effective geographical contribution to the debates calls, among other things, for 'more intensive training in analytical methods including model-based approaches, information systems and elementary analytical skills' (p. 289). What he fails to provide is either any engagement with current politics regarding the social contract that underlies taxation systems on which welfare transfers are based (and which involves both obligations and entitlements) or any discussion of norms, of what a 'good society' should have as its goals: Bennett implicitly leaves these to be determined in the market-place, and thus to be mainpulated by those with power there.

To some disciplinary historians who have adopted a periodization approach (often linked to Kuhn's paradigm model – see Johnston, 1991a), the quantitative and theoretical revolutions of the 1960s are not only over but overthrown. Later disciplinary developments are inconsistent with quantification, as Sayer (1984) argues (though see Taylor, 1981; Johnston, 1986a; Pratt, 1989). But quantification is alive and well, though not predominant, within geography and others agree with Bennett that it provides geographers with a set of skills that can be marketed in the post-welfarist world: to some those skills, especially the ones associated with the developing technology of geographical information systems (GIS), are the basis for a refocusing of 'geographical science'. In the 1960s and 1970s, quantitative skills were largely associated with deductive model-building – sometimes called spatial theory – but in the late 1980s and early 1990s, quantification has been detached from modelling by some of its main proponents. Batty (1989:156) argues intriguingly that the acquisition of quantitative skills preceded the construction of good models of the phenomena under consideration, and that 'It is an irony of history that such good models finally exist which could well have produced excellent advice in their day had they been available. But that day has passed.' Openshaw (1989) is less sanguine about the models, however: 'the deductive route to theory formulation has been taken too far and has not been particularly successful. Far too many so-called "theories" have never been tested and many more are untestable!' (p. 73) – they are 'massively theoretical' rather than applicable, and of dubious relevance.

In the welfarist world (a term he does not use), according to Openshaw:

> No-one has seemingly ever asked who are the potential users for geographical models. Previously it never mattered; today it does! Marketing geography in the world outside of academia is now becoming essential to its future survival within. The world is seemingly littered with opportunities if only geographers who had something to offer would start to look. (p. 79)

Those opportunities call for inductive rather than deductive modelling, for again according to Openshaw:

> The fundamental technical change that is underpinning the development of the new post-industrial society is the transformation of knowledge into information which can be exchanged, owned, manipulated and traded. This commodification of information in the form of computer data bases both has economic implications and creates all manner of new opportunities for models as a means of adding value to data. (p. 81)

More data exist than ever before; to get access to them, geographers will have either to pay for them in the market place or to 'join the data-keepers in providing information services' (p. 82). The opportunities to get involved in the latter are manifold, because nearly all data are, at least implicitly, locationally referenced.

Promoting geography as a discipline which adds value to spatially referenced data involves returning to the empiricist-positivist model of prediction with or without understanding: by emphasizing marketing the discipline and its technical skills, it also means focusing the provision of information on those who can pay, and thus denying it to those who can not. Openshaw describes some of the work he is involved in as producing inductive models which despite 'lacking any strong theoretical justification, have been immensely successful' (p. 82). It is possible to predict many aspects of spatial behaviour with some precision (on which, see Harvey, 1989a: 213). In doing that (Openshaw, 1989:83):

> The level of understanding is probably 'poor' in many instances but it is the best that can currently be obtained . . . In the short term, the cost of seeking applied relevancy might well be a reduction in the emphasis on explanation and the tacit acceptance of a different and inferior form of understanding . . . in essence the change amounts to no more than a switch back to an inductive paradigm to complement the *status quo* and give models in human geography a greater degree of relevance and marketability.

This data-based philosophy is supported by Rhind (1989), who writes that 'acting as a "gate-keeper" to knowledge enables one to stay "central to the action"' (p. 189). It fails to ask 'Whose action?', however – a criticism addressed to an earlier generation of 'quantitative, would-be-relevant' geographers (Harvey, 1974) – and might allow geographers to act as gatekeepers themselves, determining who obtains access to information and knowledge.

In a later polemic, defending the importance of GIS against its critics (such as Taylor, 1990), Openshaw (1991) argues that GIS 'offers the basis for a long-overdue reconciliation between the "soft" pseudoscience of social science and the "hard" spatial science of which GIS is part' (p. 621). He identifies the 'real crisis' of contemporary geography within the 'soft and the so-called intensive and squelchy-soft qualitative research paradigms' which cannot accommodate 'the new data-driven and computer-based knowledge-creating technologies' (p. 622). Geographers must come to terms with this new information-rich era, he argues, and use its tools to address the 'important geographical questions' that are waiting to be answered (p. 627), and which can provide the 'hard empirical proof' that will back up the 'realistic but not necessarily nonfictional fairy tales' that they have been weaving to account for geographical phenomena (p. 628): without that proof all that geographers have been doing is to produce 'what are no more than plausible works of fiction and then attribute to them academic credibility by developing philosophical stances that render normal scientific proof unnecessary' (p. 628). GIS offer means for 'pattern-related description and understanding' and there are few fragments of contemporary geography 'that cannot be made to fit [within it] in a fairly painless fashion and there can be few activities that will not substantially benefit as a result' (p. 628). In this view, then, geography is an empiricist science that uses information and technology to describe the world, and sells those descriptions as means to manage the world (for a reply, see Taylor and Overton, 1991).

Understanding and Emancipation

A marked contrast to the pragmatic, technocentric approach just discussed is provided by that which promotes the task of geographers as first understanding how the world is changing and then transmitting that understanding widely, as a contribution to the creation of a better world. As Harvey (1990: 418) puts it, this involves 'the construction of a historical geography of space and time', because 'societies change and grow, they are transformed from within and adapt to pressures and influences from without' (p. 419). Under capitalism, he argues, such transformation is both revolutionary and continuous:

> Capitalism is . . . a revolutionary mode of production, always restlessly searching out new organizational forms, new technologies, new lifestyles, new modalities of production and exploitation. Capitalism has also been revolutionary with respect to its objective social definitions of time and space. Indeed, when compared with almost all other forms of innovation, the radical reorganizations of space relations and of spatial representations have had an extraordinarily powerful effect . . . The elimination of spatial barriers and the struggle to 'annihilate space by time' is essential to the whole dynamic of capital accumulation and becomes particularly acute in crises of capital overaccumulation. . . . The construction and reconstruction of space relations and of the global space economy . . . has been one of the main means to permit the survival of capital into the twentieth century. (pp. 424–5)

Harvey's major (1982) theoretical exploration of the centrality of space relations to the continued capitalist dynamic was followed by his analyses of the current changes (those discussed in chapter 1 above). He argued (1989a:213) that although geographers' theoretical and quantitative revolutions have resulted in 'thousands of hypotheses proven correct at some appropriate level of significance in the geographic literature by now, . . . I am left with the impression that *in toto* this adds up to little more than the proverbial hill of beans' (see also Jackson's chapter – chapter 10 – below). The deductive modelling approach presumes 'that we know (or agree) what we are theorizing about', which for Harvey is the historical geography of space and time. Thus:

> I accept that we can now model spatial behaviours like journey-to-work, retail activity, the spread of measles epidemics, the atmospheric dispersion of pollutants, and the like, with much greater security and precision than once was the case. And I accept that this represents no mean achievement. But what can we say about the sudden explosion of third world debt in the 1970s, the remarkable push into new and seemingly quite different modes of flexible accumulation, the rise of geopolitical tensions, even the definition of key ecological problems? What more do we know about major historical-geographical transitions (the rise of capitalism, world wars, socialist revolutions, and the like)? (pp. 212–13)

He wishes to advance appreciation of the latter, within a framework which contends 'that in contemporary society commodity production for profit – i.e. capitalism – remains the basic organizing principle of economic life' (Harvey and Scott, 1989:217).

The contemporary situation is marked, according to Harvey and Scott, by a further major revolution in capitalist organization of space and time. This has been brought about by the demise of what is widely termed Fordism, characterized by mass production technologies, Keynesian economic policies, the social welfare state and 'a tripartite social contract involving the large corporations, organized labour, and the state' (p. 217: see, however, the criticisms of this view as oversimplified, as summarized by Dicken in chapter 2 above). In its place has come *flexible accumulation*:

> distinguished by a remarkable fluidity of production arrangements, labour markets, financial organization and consumption. It has at the same time engendered new rounds of what we might call time–space compression in the capitalist world – the time horizons of both private and public decision-making have shortened drastically, while electronic communications systems make it increasingly possible to spread the effects of those decisions immediately over an ever wider and more variegated space, generating major changes in patterns of geographical development . . . Organized labour, already weakened by the restructuring and high unemployment of the 1970s, has been further undercut by the reconstitution of foci of (flexible) accumulation in areas lacking in previous industrial traditions – and by the importation back into older centres of the regressive employment norms and practices established in these areas. In the new climate of entrepreneurialism, privatization and competition, the social wage has also dramatically reduced, and public austerity is now the watchword in all the advanced capitalist societies, even those still governed by socialist parties. (p. 218)

The changing world which Harvey and Scott describe is also that identified by Bennett (though in somewhat different language). They differ over how they believe geographers should respond to such changes, however,

Harvey and Scott's goal is to develop theory which will enable people to understand what is occurring, as a prelude to their altering it to create a more humane world. They:

> use 'theory' here in its Marxist sense, to mean the creation of the intellectual preconditions for self-consciousness of the structures

> of capitalist domination coupled with the construction of coherent representations and analytical tools to facilitate the struggle for human emancipation. Our ability to know the world, and to represent it truthfully, is essential to this emancipatory process. (p. 223)

Capitalism is a 'totalising system' which infiltrates most, if not all, aspects of life beyond the logic of commodity production and the operation of labour markets (they cite 'the production of information, the marketing of that information through the media, the organization of pleasure and entertainment, the production of new knowledge, the division of labour within the household' – p. 224). It comprises universals – general processes – operating 'within the fleeting, the ephemeral, the contingent, and the fragmented aspects of daily life under conditions of flexible accumulation' (p. 224). Their theoretical task is to explore the empirical world in a way which will allow them to uncover the universals, and thereby better appreciate the influences that are time-and-space contingent:

> The materialist method ... entails a search for those 'concrete abstractions' through which the capitalist mode of production (or any other mode for that matter) is bound together into a working whole. Theory construction means the conceptual representation of such concrete abstractions and their linkage into a coherent analysis through a careful reconstruction of the necessary relationships that connect them together and ensure the reproduction of capitalism as a viable social system. (p. 225)

The ultimate purpose of such a theory is emancipation, which involves geographers in a 'struggle for a different social vision and different futures with a conscious awareness of stakes and goals, albeit under conditions that are never of our own choosing' (Harvey, 1990:433). Harvey (1984) vigorously propounded this task in an earlier essay, in an analysis of the discipline's history which located its changes within the social demands of the times (as did Taylor's, 1985). To Harvey:

> Notions of 'applied' and 'relevant' geography pose questions of objectives and interests served. The selling of ourselves and the geography *we* make to the corporation is to participate directly in making *their* kind of geography, a human landscape riven with social inequality and seething geopolitical tensions. The selling of

> ourselves to government is a more ambiguous enterprise, lost in the swamp of some mythic 'public interest' in a world of chronic power imbalances and competing claims. . . . There is more to geography than the production of knowledge and personnel to be sold as commodities to the highest bidder. (p. 7)

In its place he wants a 'people's geography' which:

> reflects earthly interests, and claims, that confronts ideologies and prejudice as they really are, that faithfully mirrors the complex weave of competition, struggle, and cooperation within the shifting social and physical landscapes of the twentieth century. The world must be depicted, analyzed, and understood not as we would like it to be but as it really is, the material manifestation of human hopes and fears mediated by powerful and conflicting processes of social reproduction. (p. 7)

The goal of that geography is to 'open channels of communication, undermine parochialist world views, and confront or subvert the power of dominant classes or the state. It must penetrate the barriers to common understanding by identifying the material base to common interests' (p. 7). In doing that, he claims, geographers cannot remain neutral in how they perceive and analyse the world, but this does not prevent them from striving for scientific rigour, integrity and honesty. They must 'create an applied people's geography, unbeholden to narrow or powerful special interests, but broadly democratic in its conception' (p. 9), and which by helping others to make their own history and geography will see 'the transition from capitalism to socialism': anything else will involve them sustaining a 'present geography founded on class oppression, state domination, unnecessary material deprivation, war, and human denial' (p. 10; see Eliot Hurst's, 1985, critique of this essay and in particular the privileged place which it gives to geography, rather than the alternative 'de-disciplined' social science that he favoured).

One feature of flexible accumulation identified by Harvey and Scott is the creation of an 'ever wider and more variegated space'. The facts of spatial variation in the economic, social and political processes of restructuring have been widely recognized by geographers and others in recent years, with a resulting large literature on the study of what have become known as localities (see, for example, Cooke, 1989; Duncan and Savage, 1991). This work has been criticized by some as being little more than bland empiricism, lacking either a theoretical core (see Johnston,

1991c) or a methodological framework (Griffiths and Johnston, 1991). Harvey (1990) is critical of this approach too, while clearly accepting that flexible accumulation (Harvey and Scott, 1989:227–8):

> has encountered innumerable contingent circumstances. Its developmental course has been deeply affected by the local availability of special resources . . . , pre-existing patterns of urban and regional development, local cultures and traditions . . . It is our contention, however, that one of the crucial analytical tasks we face is to demonstrate how such contingent circumstances become internalized within, and recomposed by, the advancing development of a regime of flexible accumulation . . . we reject the idea that there is some kind of absolute opposition between theory and contingency in historical materialist analysis.

Harvey's belief in, and search for, a theory of capitalism as a totalizing system which incorporates local contingency in space and time is opposed by some, however, including those who are attracted by the arguments of postmodernism. This stresses relativism, and is 'a revolt against the too-rigid conventions of existing method and language' (Dear, 1988:265), as demonstrated, for example, by new forms of writing and of other art forms and by new architectures.

According to Cosgrove (1989a:243):

> The post-modern perspective distrusts claims for a privileged path to truth or to accurate representation of a single reality. It takes intellectual stimulus from a playful celebration of difference, from the mirroring of multiple representations. In this geography it is perhaps the theatre with its frank construction of spatial illusion, its changing scenery to indicate different meanings and moods, its dialectic of audience and players, each refracting the world of the other, and above all its languages of symbol and ritual, which offers a better metaphor for the geography of our contemporary world than the cybernetic system of macrocomputing which emerged from the flow diagrams of *Models in Geography*. The relationship between words and things has always been a contingent, socially constructed and unstable one.

Following from this, the basis of a postmodern philosophy – as argued by Dear (1988) – is a belief that claims for the superiority of one theory over another can never be decided, so that the 'intellectual conditions which allow for and tolerate the dominance of one discourse over another' (p. 265) have to be rejected. Grand theory is impossible; indeed, the search for any consensus is denigrated. The outcome is 'profoundly destabilizing and potentially anarchic' (p. 266). Instead of a grand theory one has many:

> The existence of a proliferation of social theories is one desirable quality of the postmodern approach. This approach does not aim to resolve the conflicts and contradictions between theories in favour of a single grand theory. Instead, it deliberately maintains the creative tensions between theories, in the belief that accelerated insight is likely to derive from different interpretations, revealed inconsistencies and relaxed assumptions. (p. 268)

For Dear, geography explores how the multitude of separate places comprising the world is created and recreated simultaneously (see also Soja, 1989), which no grand theory can ever account for. Harvey (1989b) disagrees, arguing that to understand the postmodern world, with its emphasis on difference rather than the conformity of modernism, requires an appreciation of the potentials created by the time–space compression of flexible accumulation. For him, the implicit nihilism of postmodernism is to be countered not by simplistic grand theorizing but rather by enhancing historical materialism in four ways (p. 355): (1) by incorporating differences as 'omni-present from the very beginning', including, for example, the roles of racial, gender and religious as well as class politics in the dialectics of social change; (2) recognizing that the production of images (works of art, architecture and literature) is part of the transformation and reproduction of any social order; (3) accepting that space and time matter; and (4) working within the belief that 'historical-geographical materialism is an open-ended and dialectical mode of enquiry rather than a closed and fixed body of understandings. Meta-theory is not a statement of total truth but an attempt to come to terms with the historical and geographical truths that characterize capitalism both in general as well as in its present phase' (p. 355). Harvey is more concerned with developing theories which can account for the existence of spatial variations, and thus give a general appreciation of the necessary geography of uneven development, than with theories which can account for specific differences: the former is possible; with the latter, the particularities of individual places are necessarily specific to them alone – and so Harvey is not promoting the type of science discussed in the next subsection.

Whereas there is clear tension between empiricist–positivist science on the one hand and realist–emancipatory science on the other, there is less conflict between the latter and humanistic science – as suggested in the book edited by Kobayashi and Mackenzie (1989). Geography, according to Cosgrove (1989a), is 'earth description' aimed at interpreting unique places, and it provides 'a sense of real people moulding their lives and their geography in real places' (p. 243). Such understanding and its transmission is not an alternative to the other types of science but rather their complement, so that whereas:

> The world remains a place of power, domination and subordination, its resources are amassed by some, denied to many, its very life processes seem fragile and threatened. All of these issues rightly bear upon the practice of geography and demand a place in its representations. But there is also beauty and goodness in this home of human life: variety, texture, colour and density, and these rest equally at the heart of geography. (p. 244)

To some, these two are neither mutually exclusive nor antagonistic: Derek Gregory (1989:v) opens his 'Foreword' to *Remaking Human Geography* by describing it as 'concerned with developing a dialogue between humanism and historical materialism' and suggests that its essays are seeking to 're-cast the relations between what some commentators see as Grand Theory (or not-so-grand theory) and "local knowledge"' (p. vi). Both humanistic and historical-materialistic geography are presented as becoming 'more participatory projects' and the essays 'demonstrate a greater sensitivity towards the deeper meaning of "making history"' (p. vii). Or, as Cosgrove (1989b:205) expresses it in the same volume:

> there is nothing to be gained by seeking some theoretical resolution to the divergent tensions within humanism and historical materialism; such a resolution would be purely rhetorical. The tension between these two ways, one of knowing and one of explaining, can be a creative one, in research to be sure, but above all in teaching our geography as a celebration of the diversity and richness of the natural world and of the places that human societies create in their attempts to make over that world into the homes of humankind.

PAME, Prediction and Planning

Some geographers clearly locate all of their discipline, both human and physical geography, within the positivist sciences. Bird, for example, accepts differences between the two main subdivisions (1989:20) – 'in physical geography testable models can approach an exactness more closely over a greater number of cases than in human geography' – but he continues by asserting that 'the ambition is the same: comprehensive exact theories to fit all cases, which is impossible'. It is impossible not only because of human fallibility but also, it seems, because of the intractability of the subject matter, especially in human geography. For that reason, Bird has developed his own methodology, based on his reading of Popper's critical rationalist philosophy.

Bird's approach – termed PAME, for Pragmatic Analytical Methodology-Epistemology – is an iterative procedure whereby putative explanations (scientific laws) are tested via 'external validation by correspondence with the real world, comparing consequences with the outputs of our work' (p. 236): this is the pragmatic element. It is analytical because it employs hypothetico-deductive procedures of reasoning rather than inductive modelling, and it proceeds in the fashion often associated with the positivist model: 'if the hypotheses are confirmed . . . happy is the researcher; if hypotheses are refuted . . . then even more interesting questions may arise as to why this is so' (p. 239: at this point, Bird is referring to the use of interview material – which he terms humanistic). Where such investigations are concerned with what he terms 'non-repetitive features of human geography' he suggests that all that can be achieved is satisfaction with the account – 'a qualitative consensus' (p. 233). But with both applied physical geography and the study of 'repetitive features of human geography (in the fields of economic and practical welfare)' (p. 232) he argues for the testing of hypotheses, while recognizing a vital difference between tests conducted to advance understanding and those used to promote a better future – 'In technology predictive success is essential, often a matter of life and death; in science we may learn more from failure of predictions' (p. 232).

Bird provides no detailed illustrations of his procedure in action. He does advocate, perhaps reluctantly, science ('knowledge for knowledge's sake') rather than technology ('applied knowledge'). Whereas he promotes the scientific approach in his statement that 'It might be best to base any criterion of quality for an academic discipline not so much on the solutions it proffers in the teaching curriculum but rather upon the problems the discipline currently identifies for the research intellects of the future' (p. 244), nevertheless he continues with a case for a bias towards technology:

> Society is placing more demands on geography. It is no use merely offering to identify problems in the physical and social environments, intellectually demanding as that might be. 'We also want solutions for our money' . . . we practice a discipline which understands where the laws of cause and effect rule in the world and where the actions of our fellow human beings, by centuries of effort to lift themselves above the sheer state of survival, have caused areas of our lives – including areas of our very spatial existence – to rise above the mere mechanical application of cause and effect; and these areas of our life may be among the most precious to us. We can offer a range of solutions based on the lessons that geography has already learnt, and be ready for the pragmatic test of the consequences of actions based on what we say.

One might expect physical geographers to support these sentiments strongly and to expect their discipline to make major technological contributions, but Haines-Young and Petch (1985) have launched a swingeing attack on the conduct of physical geography – 'there are serious problems in the way it is conducted' (p. 199) – because of the failure of its adherents to operate correctly within the accepted scientific tradition of 'theorizing, experimentation, argument and above all criticism'. Whereas practitioners in related disciplines have advanced our understanding of the natural world:

> The discipline [of physical geography], as it is taught and identified by professional physical geographers, can boast no major advances. In addition, the vast majority of journals and advanced texts still contain material which is either merely descriptive or an attempt to model some phenomenon by statistical or simple mathematical equations akin to those employed by engineers. (p. 199)

According to them, 'the engineer is more preoccupied with successful modelling and prediction than with explanation or truth' (p. 200) and they see British geomorphology 'developing into a minor branch of engineering'. They argue for adoption of the critical rationalist procedure, on which Bird's case is also built and which Hay (1985) and Marshall (1985) have commended to all geographers.

With regard to applied physical geography, Gregory (K. J. Gregory,

1985:186) has suggested that 'Physical geographers have shown a considerable reticence in becoming involved in applications of their research', but notes the growth of works published with 'applied' in their titles since the mid-1970s (see also Briggs, 1981). He categorizes such work as concerned with 'what will happen', 'how much will happen, and when' and 'how could the environment be shaped' – respectively the study of impact, prediction, and design. He expects more in the future.

Gregory's expectations appear to have been confirmed, for a review volume published just two years later (Clark, Gregory and Gurnell, 1987a) devoted six of its twenty-six chapters (98 of its 386 pages) to the theme of 'management of the physical environment'. In it, the editors accept that although physical geography remains firmly wedded to the empiricist-positivist model of science, whereas human geography also embraces humanistic and structuralist concerns, this does not necessarily imply a divorce of the two (as they believe Johnston 1983, 1986a, 1989, has argued: see also Maude, 1991): instead, 'the remainder of this book is strongly coloured by a necessary recognition that scientific techniques are to be employed in solving a menu of problems determined by humanist ideals and understood through explanations which have clear structuralist overtones' (Clark, Gregory and Gurnell, 1987b:2). Jones (1983) argued that geomorphologists had relatively neglected three areas of environmental management – policy formulation and project development; policy/project implementation; and policy and project evaluation – and the various contributors to the 1987 book both sustain that case and suggest where physical geographers ought to go, while at the same time illustrating the successes of those who have already entered the applied arena.

Newson (1989:363) argues that the core skills developed by applied physical geographers have involved 'making spatial, temporal, or joint predictions, often backed by explanations', and suggests that it might be time for the development of a greater degree of professionalism among applied physical geographers – he defines professionalism as involving 'a unified, standardised set of occupational responses to external stimuli' (p. 365). He is wary about too great a switch to applied work, however. So are Clark, Gregory and Gurnell (1987c), who suggest that 'In the 1970s the "applied" trend was so powerful that it threatened the extinction of "pure" fundamental studies' (p. 385; this statement is hard to square with the quotation from Gregory, 1985, given two paragraphs previously). They suggest that in the 1990s 'A stronger theoretical base is likely to be seen as a priority, providing a conceptual framework of greater rigour and greater flexibility than is currently available' (p. 385). Others, such as

Henderson-Sellers (1989), contend that this more rigorous framework will necessarily require greater mathematical sophistication, while Haines-Young (1989) argues that the crucial test of the models developed will be their predictive accuracy. This last criterion can only apply to work concerned with 'observable' time-scales, which excludes much physical geography. A major problem currently faced by politicians listening to geographers and others regarding the potential impacts of environmental change is that they cannot wait to see if the scientists' predictions are accurate: they must act on 'reasonable expectations' if possible disasters are to be avoided, which raises many questions as to whose 'reasonable expectations' are to be believed. The credibility of geographers as scenario painters is crucial to their potential influence. ('And is their past performance a good guide?', a question which Abler addresses in chapter 11.)

Applied geography – physical or human, or both – involves using geographers' knowledge and abilities to tackle perceived problems, either to produce a solution or to achieve a resolution (Johnston, 1986a, 1990a): it may also involve monitoring the implemented policy to evaluate its impact and, if necessary, suggest modifications. What sort of problems – which involves important issues of who defines problems, how and why – and what sort of solutions/resolutions? Berry (1981) has identified four modes of urban planning, which can be generalized to cover most geographical applications. The first involves *planning for present concerns* whereas the other three involve *planning for the future*.

1 *Ameliorative problem-solving* involves reacting to existing situations, which are themselves almost certainly consequences of past decisions. Problems are identified, interventions to tackle them are designed, and policies are put in place. Berry (p. 178) portrays this as the most common type of planning, which reflects 'the natural tendency to do nothing until problems arise or undesirable dysfunctions are perceived to exist in sufficient amounts to demand corrective or ameliorative action'. It involves simply reacting to the past and, almost certainly, haphazardly modifying the future, since the policy implemented will undoubtedly have both negative and positive feedback effects on other aspects of the system being interfered with.
2 *Allocative trend-modifying* is one of two modes that involve responding to predicted futures. Current trends are identified and ways are then found to modify them so as to avoid potential future problems and make the best of the trends.
3 *Exploitative opportunity-seeking* is similar to the previous type except

that it involves no attempts to modify trends: the goal is to make the most of what is likely to happen, without being too concerned about problems that may arise as a consequence.

4 *Normative goal-oriented* planning differs from the other three in that it starts not by identifying current problems and/or trends and using them as the base for policy design but rather by determining the desired future and only then creating policies which will ensure that it is attained.

Berry (1981: 179) points out that:

> The four different planning styles have significantly different long-range results, ranging from haphazard modification of the future produced by reactive problem-solving, through gentle modification of trends by regulatory procedures to enhance existing values, to significant unbalancing changes introduced by entrepreneurial profit-seeking, to creation of a desired future specified *ex ante*. Clearly, in any country there is bound to be some mixture of all styles present, but equally, predominant value systems so determine the preferred policy-making and planning style that significantly different processes assume key roles in determining the future in societies.

In this context, an important issue for geographers is the degree to which they accept a role that involves them in what may be interpreted as imposing a future on others. Buttimer (1979:30) argues that such applied geography is explicitly managerial, involving the manipulation of people and their environments rather than contributing to an individual's process of 'human becoming'. For her (Buttimer, 1974:29), 'the social scientist's role is neither to choose or decide for people, nor even to formulate the alternatives for choice but rather, through the models of his discipline, to enlarge their horizons of consciousness to the point where both the articulation of alternatives and the choice of direction could be theirs'.

For some, that goal – however laudable in principle – may be unattainable, either because of the specialist knowledge needed to appreciate the nature of both current problems and likely futures or because of the need for collective rather than individual decision-making, which invariably involves domination by those with political power (and of varying degrees of accountability). Such writers claim that it is for the specialists to make clear what is happening and what might happen, and to encourage those with power to plan accordingly: to do otherwise would be to abdicate the responsibility which their knowledge gives them. Thus in determining what type of applied geography, and so what type of science, they wish to practise, geographers must decide whether they wish to set themselves up either as experts who, at best, help people decide what is best for them (and, at worst, present themselves as knowing what is best and thus deserving the power to implement that knowledge), or as analysts who assist people to know themselves and the societies in which they live, so that they can better evaluate their presents and control their own futures.

What Problems?

The challenges are many, as are the possible responses. Should geographers concentrate their research effort (as Clayton, 1985, suggested, for political reasons: see also chapter 11), or should they be prepared to tackle whatever issues attract their attention? Stoddart has claimed that the main difficulty faced in framing an answer to that question is that geographers have lost their *raison d'être* within science. The introduction to his book of essays on the discipline's history begins with the claim that he is a geographer working in a great tradition, but (Stoddart, 1986:ix):

> I recognize – and regret – the fact that for many people the geographer has long since appeared to have surrendered to other specialists his catholic concern for the diversity of the natural world. We have too long accepted the artificial constructions of the bookmen about what geography is, what it should be concerned with and how it should be done. I confess to a feeling of unreality about much of the literature on the philosophy, methodology and even history of the subject, most of it written by people who signally fail to practise what they preach. Meanwhile so many retreat into increasingly restrictive and esoteric specialities, where they protect themselves with secret languages and erudite techniques.

He finds that his contemporaries as geographers 'on the human and quantitative sides of the subject' suffer from doubts and despair about their discipline, but puts this down to the sorts of things that they study – 'I cannot be surprised that "hours of thoughtless number crunching" on things like the distribution of gasworks in Lancashire or service stations on motorways lead to this kind of moroseness' (p. x) – and gives them the simple advice 'do some *real* geography' which occupies 'a central role among the sciences of nature' (p. x). Stoddart (1987) expanded on this theme a year later, using the Carl O. Sauer Memorial Lecture at the University of California, Berkeley, to proclaim that 'Geography is alive and well, that the disaffected speak only for themselves' (p. 329). They are disaffected, he argues, because 'the intellectual horizon that these people recognize as Geography is to me no longer recognizable as Geography at all' (p. 329). They have divided and subdivided the discipline into a series of specialisms each with 'its own expertise, . . . [its] own techniques, . . . [its] own theoretical constructs', so that 'Across geography we speak separate languages, do very different things. Many have abandoned the possibility of communicating with colleagues working not only in the same titular discipline but also in the same department' (p. 330). (For a further perspective on such fragmentation, see Johnston, 1991c.) Within many of the specialisms, geographers study topics that for Stoddart are worse than trivial (as Goudie has quoted in chapter 6, but it bears repeating):

> Quite frankly . . . I cannot take seriously those who promote as topics worthy of research subjects like geographic influences in the Canadian cinema, or the distribution of fast-food outlets in Tel Aviv. Nor have I a great deal more time for what I can only call the chauvinist self-indulgence of our contemporary obsession with the minutiae of our own affluent and urbanized society – housing finance, voting patterns, government subsidies for this and that, and how to get most from them. We cannot afford the luxury of putting so much energy into peripheral things. Fiddle if you will, but at least be aware that Rome is burning all the while. (p. 334)

The clue to Stoddart's alternative is given in the final sentence of the last quotation. He writes approvingly of earlier generations of geographers who 'dared to do something that we, in our sophistication, rarely do. They asked the big questions, about man, land, resources, human potential. . . . We need to remember that science is about asking daring questions like these' (p. 334). Unfortunately, he contends:

> We no longer ask these questions, but the questions remain. It is largely people other than geographers who are asking – and answering – them now. It is astonishing that it is Ladurie and the *Annales* school who have commandeered the whole field of the relations of climate and history. Braudel writes what is in effect geography (though without maps) and calls it history: the historical geographers tag along in dutiful homage.

Geographers must reclaim their intellectual heritage:

> We need to claim the high ground back: to tackle the real problems: to take the broader view: to speak out across our subject boundaries on the great issues of the day (by which I do not mean the evanescent politics of Thatcher, Reagan and Gorbachov). We need to forget the trivia and the tedium of much that has passed for geographical research and erudition over the past twenty years ... Land and life is what geography has always been about. It is time we got out again into the great wide world, met its challenges, and met them in a way that Forster, Humboldt and Carl Sauer himself would have approved. (p. 333)

Bird (1989) pointed out that Stoddart gives no clear statement as to the geographical methodology to be used in the reclamation of the high ground. Stoddart uses the current problems of Bangladesh to illustrate the sorts of issues that geographers should be tackling – problems that are specific to a region and reflect the interrelationships of society and environment there. It is, Stoddart claims, 'real, unified and committed' (p. 333) and involves both physical and human geography: 'there is no such thing as a physical geography of Bangladesh divorced from its human geography, and even more so the other way round. A human geography divorced from its physical environment would be simply meaningless nonsense' (p. 333). His alternative would also be 'well worth while', for geography 'must be a subject which provides a means of creating feelings worthy of humanity; it must fight against racialism, war, intolerance and oppression; it must disperse the lies resulting from ignorance, presumption and egotism' (p. 333). In some ways, Stoddart's geography sounds like that advocated by Harvey (see above, p. 163):

> It is a geography which reaches out to the future, and the future is even conditional on how well we do it. It is a geography which will teach us the realities of the world in which we live, how we can live better on it and with each other. It is a geography which will teach our neighbours and students and our children how to understand and respect our diverse terrestrial inheritance. (Stoddart, 1987:333)

But Stoddart differs from Harvey more than he agrees with him, for he is defending a disciplinary tradition based (as Taylor and Jackson argue below, in chapters 9 and 10), inter alia, on ethnocentrism, imperialism and masculinism (see Stoddart's, 1991, response to Domosh, 1991a, and also Domosh, 1991b): Harvey is not, for he is forward- rather than backward-looking in his search for a disciplinary role.

In apparently restricting the discipline's subject matter to society–environment interrelationships, Stoddart precludes development of the understanding necessary to fight racialism, wars, intolerance and oppression and to counter the lies resulting from ignorance, presumption and egotism. One can readily accept his goal, but not his prescribed route there.

Nevertheless, some others, such as Bennett (1989:281), apparently agree that we are evading the major issues which involve the interactions between society and the physical environment and demeaning our discipline accordingly:

> Geography, within this [current] conception of 'social science', loses its distinctive disciplinary focus either as the study of the relations of environment and society, or as the study of absolute and relative space; both of these traditional focuses include major 'non-social' aspects. The alternative social science definition de-defines significant parts of the discipline as it has normally been conceived.

But what are those relationships, and how should they be investigated by scholars drawing on the separate (and perhaps incommensurate) philosophies of the natural and social sciences (Johnston, 1989)? Too many geographers concerned with the 'big issues' relating to the environment are either unprepared to confront the understanding of society provided by social science, without which the interrelationships cannot be appreciated, or (like Bennett, it seems) have a very myopic view of social science.

WHAT SHOULD WE DO?

The four cases for particular forms of geography outlined above have been presented as exemplars of the sorts of argument currently alive within the discipline. They were not selected to give either a comprehensive or a representative survey, but rather to provide a flavour of some of the more provocative recent writings. All of them argue for change in the practice of geography, and thus to some extent imply a contemporary malaise within the discipline. By drawing them out, this chapter has hopefully assisted others in their quest for a satisfying role within a changing world: the goal has been emancipatory, not prescriptive.

Clearly the arguments could be assessed further, debate could be engaged, and further points of view assessed. The following three chapters in part do that. They, too, are neither representative nor comprehensive in the topics covered and the proposals made. Rather they represent the views of other geographers concerned to enhance not only the particular role of the discipline but also the general contribution of the 'knowledge industry' to facing the challenge of a changing world. They ask questions of us all.

The chapters by Peter Taylor (9) and Peter Jackson (10) both stress the need for geographers to recognize the position from which they speak, but in very different ways. Taylor develops his argument by revisiting the discipline's twentieth-century history, and in so doing presenting similar material to Stoddart's (1986), but from a very different standpoint. One hundred years ago, geography was very much 'a European science' (a point also made by Stoddart, 1982, though in a quite different way): the global geographies being written then were extremely cultural-specific. Following a brief flirtation with the quantitative social sciences, geographers are returning to the global view. This will necessarily be very different from the first version, however, not just because the world has changed radically over the century but also because the Eurocentrism of the first global geographies is no longer acceptable. If geography is to be an emancipatory discipline, as Harvey and others have argued, then it must be based on an empathy for the differences that the global mosaic comprises, not on an assumed superiority of one viewpoint (on which see the important book by Said, 1978, which has clearly influenced both Taylor and Jackson).

Jackson, by contrast, draws his inspiration from feminist and anthropological sources, which stress the need for understanding the positionality of a scholar (either oneself or one whose works are being read) – we write from within a context, and what we write can only be understood if one has an appreciation of that context, so that the futures that we promote are context-dependent too. The world that we inhabit is a world of containers, some that we make for ourselves but most made for us by others. Those containers have boundaries, and much political struggle is over either their movement or their demolition, though many are unaware of the struggles and of the arguments being developed outside the containers that they themselves occupy. We need to get inside those containers – metaphorically if not literally – in order to appreciate the great range of positions from which people are speaking: in doing so, we will change ourselves, and contribute to the changing of others.

The applied geographies argued for by Taylor and Jackson are unlikely to find favour with the critics who castigate human geographers for their 'soft pseudoscience' and 'squelchy-soft qualitative research'. Nor do they seem to fit into the type of applied geography which Taylor's (1985) analysis sees as ascendant at present: a decade and a half ago, Harries (1976) – responding to Peet's (1975) critique of 'geographies of crime' – argued that being a radical was a luxury few academics could afford, and that working at a publicly funded American university demanded a pragmatic rather than a revolutionary attitude! The increasingly materialistic view taken of higher education in many countries in recent years has stimulated many geographers, among others, to consider the orientation of both their research and their teaching. Ron Abler suggests (chapter 11) the directions that geographers might take, emphasizing the practical. In a statement that paraphrases John F. Kennedy's famous call in his inaugural address, he invites us to ask not what society can do for geographers but rather what geographers can do for society. This will involve agenda-setting at an institutional scale, will probably require specialization among institutions, and will need geographers to cooperate much more in large research teams than has been the case heretofore.

So which route do we follow? Do we emphasize technique and the attack on practical issues, large and small, or do we seek to educate in the true sense of that term, which is what Jackson and Taylor commend? Or are we large enough, strong enough and diverse enough to succeed at both? A positive response to the last question may suggest a misguided complacency: the world is changing fast, so how should geographers change with and for it?

NOTES

1 These items are occasionally vituperative and involve conflicts among geographers in the 'public arena', as with the essay by David (1958), 'Against geogaphy'. For a contemporary example of the same genre, see the letter by W. H. Parker in the English newspaper the *Daily Telegraph* for 22 January 1992.

REFERENCES

Batty, M. 1989: Urban modelling and planning: reflections, retrodictions and prescriptions. In B. Macmillan (ed.), *Remodelling Geography*, Oxford: Basil Blackwell, 147–69.

Bennett, R. J. 1989: Whither models and geography in a post-welfarist world? In B. Macmillan (ed.), *Remodelling Geography*, Oxford: Basil Blackwell, 273–90.

Bennett, R. J. (ed.), 1990a: *Decentralization, Local Governments and Markets: towards a post-welfare agenda*. Oxford: Clarendon Press.

Bennett, R. J. 1990b: Decentralization, intergovernmental relations and markets: towards a post-welfare agenda. In R. J. Bennett (ed.), *Decentralization, Local Governments and Markets: towards a post-welfare agenda*, Oxford: Clarendon Press, 1–28.

Berry, B. J. L. 1981: *Comparative Urbanization*. New York: St Martin's Press.

Bird, J. H. 1989: *The Changing Worlds of Geography*. Oxford: Clarendon Press.

Briggs, D. J. 1981: The principles and practices of applied geography. *Applied Geography*, 1, 1–8.

Buttimer, A. 1974: *Values in Geography*. Commission on College Geography, Resource Paper 24, Association of American Geographers, Washington, DC.

Buttimer, A. 1979: Erewhon or nowhere land. In S. Gale and G. Olsson (eds), *Philosophy in Geography*, Reidel: Dordrecht, 9–38.

Clark, M. J., Gregory, K. J. and Gurnell, A. M. (eds), 1987a: *Horizons in Physical Geography*. London: Macmillan.

Clark, M. J., Gregory, K. J. and Gurnell, A. M. 1987b: Introduction: change and continuity in physical geography. In M. J. Clark, K. J. Gregory and A. M. Gurnell (eds), *Horizons in Physical Geography*, London: Macmillan, 1–5.

Clark, M. J., Gregory, K. J. and Gurnell, A. M. 1987c: Physical geography: diversity and unity. In M. J. Clark, K. J. Gregory and A. M. Gurnell (eds), *Horizons in Physical Geography*, London: Macmillan, 382–6.

Clayton, K. M. 1985: The state of geography. *Transactions, Institute of British Geographers*, NS 10, 5–16.

Cooke, P. N. (ed.) 1989: *Localities*: London: Unwin Hyman.

Cosgrove, D. E. 1989a: Models, description and imagination in geography. In B. Macmillan (ed.), *Remodelling Geography*, Oxford: Basil Blackwell, 230–44.

Cosgrove, D. E. 1989b: Historical considerations of humanism, historical materialism and geography. In A. Kobayashi and S. Mackenzie (eds), *Remaking Human Geography*, Boston, MA: Unwin Hyman, 189–205.

David, T. 1958: Against geography. *University Quarterly*, 12, 261–73.

Dear, M. J. 1988: The postmodern challenge: reconstructing human geography. *Transactions, Institute of British Geographers*, NS 13, 262–74.

Domosh, M. 1991a: Towards a feminist historiography of geography. *Transactions, Institute of British Geographers*, NS 16, 95–104.

Domosh, M. 1991b: Beyond the frontiers of geographical knowledge. *Transactions, Institute of British Geographers*, NS 16, 488–90.

Duncan, S. S. and Savage, M. 1991: New perspectives on the localities debate. *Environment and Planning A*, 23, 155–64.

Eliot Hurst, M. E. 1985: Geography has neither existence nor future. In R. J. Johnston (ed), *The Future of Geography*, London: Methuen, 59–91.

Entrikin, J. N. 1991: *The Betweenness of Place*. London: Macmillan.

Gregory, D. 1978: *Ideology, Science and Human Geography*. London: Hutchinson.

Gregory, D. 1980: The ideology of control: systems theory and geography. *Tijdschrift voor Economische en Sociale Geografie*, 71, 327–42.

Gregory, D. 1989: Foreword. In A. Kobayashi and S. Mackenzie (eds), *Remaking Human Geography*, Boston, MA: Unwin Hyman, v–viii.

Gregory, K. J. 1985: *The Nature of Physical Geography*. London: Edward Arnold.

Griffiths, M. J. and Johnston, R. J. 1991: What's in a place? *Antipode*, 23, 185–213.

Habermas, J. 1972: *Knowledge and Human Interests*. London: Heinemann.

Haines-Young, R. 1989: Modelling geographical knowledge. In B. Macmillan (ed.), *Remodelling Geography*, Oxford: Basil Blackwell, 22–39.

Haines-Young, R. and Petch, J. R. 1985: *Physical Geography: its nature and methods*. London: Harper and Row.

Harries, K. D. 1976: Observations on radical versus liberal theories of crime causation. *The Professional Geographer*, 28, 100–13.

Harvey, D. 1974: What kind of geography for what kind of public policy?. *Transactions, Institute of British Geographers*, 63, 18–24.

Harvey, D. 1982: *The Limits to Capital*. Oxford: Basil Blackwell.

Harvey, D. 1984: On the history and present condition of geography: an historical materialist manifesto. *The Professional Geographer*, 36, 1–11.

Harvey, D. 1989a: From models to Marx: notes on the project to 'remodel' human geography. In B. Macmillan (ed.), *Remodelling Geography*, Oxford: Basil Blackwell, 211–16.

Harvey, D. 1989b: *The Condition of Postmodernity*. Oxford: Basil Blackwell.

Harvey, D. 1990: Between space and time: reflections on the geographical imagination. *Annals of the Association of American Geographers*, 80, 418–34.

Harvey, D. and Scott, A. J. 1989: The practice of human geography: theoretical and empirical specificity in the transition from Fordism to flexible accumulation. In B. Macmillan (ed.), *Remodelling Geography*, Oxford: Basil Blackwell, 217–29.

Hay, A. M. 1985: Scientific method in geography. In R. J. Johnston (ed.), *The Future of Geography*, London: Methuen, 129–42.

Henderson-Sellers, A. 1989: Climate, models, and geography. In B. Macmillan (ed.), *Remodelling Geography*, Oxford: Basil Blackwell, 117–46.

Johnston, R. J. 1983: Resource analysis, resource management, and the integration of human and physical geography. *Progress in Physical Geography*, 7, 127–46.

Johnston, R. J. 1986a: *On Human Geography*. Oxford: Basil Blackwell.

Johnston, R. J. 1986b: Perspectives on applied human geography. In J. Bosque Maurel (ed.), *Homenage a Manuel de Teran Alvarez*, Madrid: Universidad Complutense, 125–40.

Jonhston, R. J. 1989: *Environmental Problems: nature, economy and the state*. London: Belhaven Press.

Johnston, R. J. 1990a: Some misconceptions about conceptual issues. *Tijdschrift voor Economische en Sociale Geografie*, 81, 14–18.

Johnston, R. J. 1990b: Exploring graduate dissatisfaction with British geography degree courses. *Journal of Geography in Higher Education*, 14, 39–54.

Johnston, R. J. 1991a: *Geography and Geographers: Anglo-American human geography since 1945* (fourth edition). London: Edward Arnold.

Johnston, R. J. 1991b: Graduate evaluation of British higher education courses. *Studies in Higher Education*, 16, 209–24.

Johnston, R. J. 1991c: *A Question of Place*. Oxford: Basil Blackwell.

Jones, D. K. C. 1983: Environments of concern. *Transactions, Institute of British Geographers*, NS 8, 429–57.

Kobayashi, A. and Mackenzie, S. (eds), 1989: *Remaking Human Geography*. Boston, MA: Unwin Hyman.

Marshall, J. U. 1985: Geography as a scientific enterprise. In R. J. Johnston (ed.), *The Future of Geography*. London: Methuen, 113–28.

Maude, A. 1991: Integrating human and physical geography? Teaching a first year course in environmental geography. *Journal of Geography in Higher Education*, 15, 113–22.

Newson, M. D. 1987: From description to prescription: measurements and management. In M. J. Clark, K. J. Gregory and A. M. Gurnell (eds), *Horizons in Physical Geography*, London: Macmillan, 353–68.

Openshaw, S. 1989: Computer modelling in human geography. In B. Macmillan (ed.), *Remodelling Geography*, Oxford: Basil Blackwell, 70–88.

Openshaw, S. 1991: A view on the GIS crisis in geography, or, using GIS to put Humpty-Dumpty back together again. *Environment and Planning A*, 23, 621–8.

Pacione, M. 1990: Conceptual issues in applied urban geography. *Tijdschrift voor Economische en Sociale Geografie*, 81, 3–13.

Peet, J. R. 1975: The geography of crime: a political critique. *The Professional Geographer*, 27, 277–80.

Pratt, G. 1989: Quantitative techniques and humanistic-historical materialist perspectives. In A. Kobayashi and S. Mackenzie (eds), *Remaking Human Geography*, Boston, MA: Unwin Hyman, 101–15.

Rhind, D. W. 1989: Computing, academic geography and the world outside. In B. Macmillan (ed.), *Remodelling Geography*, Oxford: Basil Blackwell, 177–90.

Said, E. 1978: *Orientalism*. New York. Pantheon Books.

Sayer, A. 1984: *Method in Social Science*. London: Hutchinson.

Soja, E. W. 1989: *Postmodern Geographies: the reassertion of space in critical theory*. London: Verso.

Stoddart, D. R. 1982: Geography – a European science. *Geography*, 67, 289–96.

Stoddart, D. R. 1986: *On Geography: and its history*. Oxford: Basil Blackwell.

Stoddart, D. R. 1987: To claim the high ground: geography for the end of the century. *Transactions, Institute of British Geographers*, NS 12, 327–36.

Stoddart, D. R. 1991: Do we need a feminist historiography of geography – and if we do, what should it be?. *Transactions, Institute of British Geographers*, NS 16, 484–7.

Taylor, P. J. 1981: Factor analysis in geographical research. In R. J. Bennett (ed.) *European Progress in Spatial Analysis*, London. Pion, 251–67.

Taylor, P. J. 1985: The value of a geographical perspective. In R. J. Johnston (ed.), *The Future of Geography*, London: Methuen, 92–110.

Taylor, P. J. 1990: Editorial comment: GKS. *Political Geography Quarterly*, 9, 211–12.

Taylor, P. J. and Overton, M. 1991: Further thoughts on geography and GIS. *Environment and Planning A*, 23, 1087–90.

9

Full Circle, or New Meaning for the Global?

Peter J. Taylor

> It seems that the world is making a comeback into British geographical thought . . . It is perplexing, of course, that the world should ever have disappeared from view, but now that it has returned let us welcome it wholeheartedly.
>
> Alan Gilbert, 1987: 130

'The pursuit of the global view', so Stephen B. Jones (1954:159) tells us, 'is the geographer's intellectual adventure.' No single statement could sum up better the basis of my particular romance with Geography.[1] Geography is about the world and we should never forget it. But of course many, perhaps a majority, of geographers have never felt the need to pursue a global view as they become bogged down by their particular empirical minutiae or abstract theory. Hence the common concern that somewhere in its history Geography lost its 'geo'. Ironically, this was never more the case than at the height of the Cold War, when Jones was writing. With few exceptions, when the world as the home of humanity was threatened as never before, geographers retreated to studies at the local or state scale. There remains very little written in Geography on either the Cold War geopolitical world order or the nuclear shadow that was hanging over it. Hence Stephen Jones, like many political geographers before and since, was 'out of scale' from the rest of the discipline (Claval, 1984).

It has not always been so. Political geographers and geographers in general have been equally concerned with global issues at certain times. We may well be entering just such a time. For instance, Ron Johnston

(1985:325) has castigated what he terms Geography's 'disengagement from the world' as 'myopic' and 'parochial'. He goes on to plead with contemporary geographers to 'return' to the world and present it 'in its full diversity' as an educational vehicle for 'international awareness and peace' (p. 334). Johnston's point is simply that whereas geographical knowledge of the world may promote world citizenship, geographical ignorance of the world may lead to xenophobia. Hence his call for Geography to fulfil its 'traditional role' to promote world understanding between peoples (p. 328). In effect he sees Geography turning full circle to embrace the global scale of analysis that had been so important at the beginning of this century in the work of such prominent geographers as Sir Halford Mackinder. Unstead and Taylor (1910:1), for instance, defined Geography at that time as 'the science which deals with the distribution of various phenomena over the Globe'. Nevertheless, we must be careful about making comparisons between Geographies devised in very different contexts. To put it another way, Mackinder's 'intellectual adventure' is very different from Johnston's, despite their common concern for the global.

The global view adopted by geographers at the beginning of the century was very much set in an imperialist context. In this chapter I will describe this particular project as a geographical monologue wherein Europe dominates and designates the non-European world. In contrast, the global view that Johnston is promoting entails a geographical dialogue between the different regions of the world that transcends past Eurocentrism. Hence it is problematic, to say the least, whether Johnston can appeal to traditional Geography to legitimate his return to the global scale. He is embarking on a quite different project: instead of a full circle, what we are offered is a new meaning for the global in geography.

The purpose of this essay is to illustrate this argument by describing the two 'global Geographies' that have existed this century. The first three subsections describe the first of these Geographies in terms of its imperialist context, its codification as a discipline and its practices in writing global Geography. We then have an interlude where we consider briefly the Geography that lost its 'geo'. In the final three subsections we present what a global Geography might look like once it sheds its imperialist garb. As a geographical dialogue the discipline can become a truly world subject.

GLOBAL GEOGRAPHY I

Geographical Monologue: European Subject, World Object

The origins of modern Geography are inexorably tied to the practice of exploration. Not any exploration, of course, but the phase beginning in the 'long sixteenth century' whereby Europeans created the global scale of activities that are our concern here. Before our century the terms 'geography' and 'exploration' were largely interchangeable. For instance, before Geography became a university discipline at the end of the nineteenth century, histories of Geography mostly consisted of descriptions of exploration (Grano, 1980). More generally, geographical societies were, above all else, exploration societies. It is hardly surprising, therefore, that David Stoddart (1982) has been able to use this exploration heritage to designate Geography 'a European science'. He concludes his discussion: 'in method and in concept Geography, as we know it today, is overwhelmingly a European discipline, which emerged as Europe encountered the rest of the world' (p. 295). The use of the verb 'encounter' is interesting in this context, since it implies a rather neutral interaction between Europeans and non-Europeans. The most common verb used to describe this activity has been 'to discover'. William Bunge (1965) exposed many years ago the racial arrogance behind the treatment of exploration as discovery – 'how backward of them for us not to have found them' (p. 1): it seems that non-European peoples were not actually lost before 'we discovered them'! He ridicules European 'discovery' by relating the story of the Royal Ethiopian Explorers Society, which was particularly fascinated by 'the problem of the Rhine'. In a similar vein, Australian indigenous people celebrated the two hundredth anniversary of Australia in 1988 by landing on a beach in England to 'discover' and lay claim to Britain. The point is well made; many other worlds existed before the Europeans created our contemporary global society. Following Bunge, therefore, we should interpret the explorations since 1492 as extending the known world of Europeans to its global limits.

The past popularity of the verb 'to discover' reflects the power differential in the European exploration project: it is the strong who discover, explore and travel; the weak are discovered, explored and visited. It is a European world that has been produced over the last five hundred years. Hence the geography created in its wake does not simply reflect an encounter between Europe and the rest of the world, it is integral to the European domination of the world. One way of expressing this is to view it as a geographical monologue; Europe told the rest who and what they were in its new world. In Rana Kabbani's (1988) apt phrase, Europe's domination included the process of 'devise and rule'. The best description of this process can be found in Edward Said's (1978) investigation of Orientalism: Asia 'was not found to be "Oriental"' rather it was 'orientalized' as part of the European incorporation of Asia into its world.

Eric Wolf (1982) has described the non-Europeans before their 'discovery' as 'people without history' in the sense that their story is deemed to begin only after their integration into our world. Similarly Frederick Turner (1980) describes the regions outside the European known world as 'beyond geography'. Hence the Europeans gave to the peoples of the rest of the world not just their history but their geography also. We may say, therefore, by rephrasing Lenin, that Geography became the study of the world as described by the winners. This is very obvious in the names by which the major regions of the world are commonly known: Middle East, Far East and West Indies, for example, all reflect a Eurocentric view of the world. In fact it is only in recent years that this terminology has been challenged – *Third World Guide* (Bissio, 1988) replaces Middle East by Majrek, for instance, and Aprismo politics has long preferred Indoamerica as an anti-imperial alternative to Latin America (Alexander, 1973:357). This movement away from a European-devised world is perhaps most apparent in the recent debates on global map projections. Whatever criticisms we may have of the Peters projection, it has done Geography the great service of opening up a longoverdue debate on how our world should be depicted (Vujakovic, 1989). But for most of the history of our modern world-system, there was little or no such concern for rampant Eurocentrism in Geography. This European science devised the world that Europeans ruled. It was both blind and deaf to the worlds of others. It was a geographical monologue.

Making a Discipline at the Fin de Siècle

In the late nineteenth century, geography began the task of freeing itself from its exploration heritage with the new project of constituting itself as a university discipline. This was the era of discipline-making in Europe as universities were expanded to provide new, highly qualified labour for the more complex societies and states that were emerging. Intellectually the expansion produced a very fluid situation, with many claims for studies to be accepted as university subjects. Whereas History could continue in its traditional narrative mode, safe in its sole 'possession' of the human written past, Geography's concern for the contemporary was challenged by the new social sciences. Hence Geography struggled to be accepted as a university discipline. As a simple description of the earth's surface, its intellectual credentials were doubted. In this view, Geography should be considered a preliminary subject that merely provides the platform for historians and social scientists, geologists and biologists to develop their more advanced explanations and theories. It is against this background that the project of making Geography a modern discipline had to be pursued.

In order to creat a new discipline Geography needed a distinctive theoretical base, its own object to study and an acceptable methodology to pursue its aims. These were all achieved in the decades either side of 1900. Geography dismissed its image as an inert platform by encompassing the social theory of environmentalism. Debates raged about the form that environmental influences on society took, but the content of Geography in terms of society–environment relations was not disputed. These relations were expressed in different ways in different places, and hence the region could be designated as the object of geographical study. These homogeneous zones of environmental influence could vary in scale from global 'natural regions' to small French '*pays*', but they were all deemed to express the same general environmentalism. Finally, a methodology evolved that matched the needs of the new discipline. Its distinctive environmental basis meant that its subject matter straddled the divide between physical and human sciences. Geographers made a virtue out of this problem by adopting synthesis as their fundamental method to bridge the divide. Hence regional synthesis became the ultimate 'art' of the geographer. Others might analyse their particular subjects but it was the geographer's task to bring it all together in his or her regional geography.

In hindsight, the success of the early geographers in establishing their discipline within universities is quite surprising. The Geography that was created developed ideas that ran counter to the prevailing intellectual trends (P. J. Taylor, 1985). As universities grew, new disciplines had been constructed out of narrower and narrower specialisms. Geography, on the other hand, broadly encompassed a vast array of material. And while others were refining their tools of analysis, geographers were proclaiming their art of regional description. No wonder Fred K. Schaefer (1953) was to castigate this Geography as exceptionalist and attempt to bring it back into the scientific and intellectual mainstream.

There was one exceptional feature of this Geography that is less commented upon. The social sciences that were created at this time took as their unexamined basis the state as their object of study. Economies, societies and polities were all equated with activities taking place within state boundaries. Economics, Sociology and Political Science were devised as creatures of the state. This form of state-centric thinking was modified in Geography. The choice of region as object of study provided an alternative to the state as the concrete entity of interest. Regions could be either much larger than the state or much smaller, and, further, their boundaries could cut across state boundaries. This is not to say that geographers were not interested in states, especially their own state – Mackinder's biographer quite reasonably subtitles his study 'geography as an aid to statecraft' (Parker, 1982) – but that they provided a different spatial lens for viewing the world largely neglected by the social sciences. Regions smaller than the state could be useful for internal domestic planning; regions larger than the state could be of value as a framework for foreign policy. And this was important in the political context of the times, where acute international rivalries focused upon issues such as autarchy and the imperial harvest.

What Did a Global Geography Look Like?

It has become generally accepted that we can only understand the history of geography within the social context of its production and reproduction (Berdoulay, 1980). At the beginning of this century, that context can be summarized in just one word – imperialism. Hudson's (1977) seminal paper has illustrated the way in which Geography was a child of imperialism. Perhaps the most explicit intellectual link can be found in Herbertson's (1910:477) discussion of 'the need of recognizing the Geographical factor in Imperial Problems'. He proposed that the natural regions of the world be mapped in terms of their current economic value so that they could be developed into 'other maps prophetic of economic possibilities'. Such imperial planning would be the responsibility of 'Geographical-Statistical Departments' staffed by geographers to advise government. Although such ambitious plans never materialized, the close links between Geography and imperialism are not in doubt and provide a very strong 'external' reason for the success of Geography in establishing itself as a discipline (P. J. Taylor, 1985).

Geography's relationship with imperialism is much more than as an actual or potential instrument for imperialist ends, however. The whole period of producing Geography as a European science is imbued with assumptions of imperialism. Sachs (1976) calls this 'the Vasco de Gama era', which lasts from approximately 1500 to 1950. As new worlds were discovered and old worlds reinterpreted, a dominant world view emerged of a white, civilized centre and a barbaric, subjugated periphery of non-white natives. According to Sachs (p. ix), this view 'collapsed beneath the wave of emancipation' in the decade following World War II. The geographers working in and for the great European imperialist states undoubtably subscribed to this general world view, but so too did many geographers who would consider themselves to be anti-imperialist. In the United States, for instance, the tradition of anti-European imperialism did not prevent the major geographers of the day accepting the same global hierarchy of peoples as the Europeans *and* with the same policy implications. Derwent Whittlesey (1939) discusses the 'White Man's Burden' in his classic political geography text, and provides us with a world map of 'The Exploitable World' which consists of six regions – Africa, Antarctica, the Levant, Middle America, the Orient and the Pacific. There seems little doubt who is exploiting whom in this map. Two years later the other leading political geographer of that time, Richard Hartshorne, produced another world map, this one entitled 'Effective Political Organization' (Bergman, 1975:19). On it nearly all of Africa, most of South America and much of Asia is designated 'permanently colonial'. Clearly Hartshorne did not think the peoples of these areas capable of running their own affairs: he was a typical thinker of the Vasco de Gama era.

Perhaps the most complete expression of this 'Vasco de Gama Geography' can be found as the era comes to an end, in Griffith Taylor's (1951) collection of essays *Geography in the Twentieth Century* and J. H. G. Lebon's (1951; in references as Lebon, 1963) little textbook *An Introduction to Human Geography*. Taylor used his edited volume to defend his own geographical determinism, but the most notable feature is the widespread concern with global issues: 'After all, Geography is the science of the *world*: and it is the understanding of the whole world that concerns us mainly' (p. 9, italics in original). This volume can be interpreted as the final swan-song of the first global Geography. Taylor introduced a 'geopacifics' to replace discredited geopolitics, and foresaw 'settlement forecasting' as a major area of growth in future Geography (p. 4). Neither new subdiscipline really outlasted the book. The latter represented Geography's traditional fascination with frontiers, which is well-represented in the volume. In the chapter on 'Geography and the Tropics', for example, Karl J. Pelzer (1951:315–16) treated the subject in an extreme Eurocentric manner:

> Political as well as professional and lay circles in Europe have great hope that the African tropics may prove to be the panacea for many

> of the ills which harass Europe today . . . the interests of the indigenous population have to be fully protected; some even go as far as to demand that they should come first.

The last sentence provided hints of a new era, but not much more. The remainder of the discussion was about the labour problem in the tropics and the question of white acclimatization.

Perhaps the most characteristic world map epitomizing this global Geography was one showing 'The Limits of Settlement Possibilities' which highlights the 25 °C mean annual temperature isotherm (between pages 264 and 265). This map ignores the peoples already living in the tropics – India with its hundreds of millions of people is located beyond the limits of settlement possibilities! – and is concerned only with external settlers. This first global Geography remains a European monologue to the very end.

This 'practical imperialism' was also very clear in Lebon's very popular textbook, which went into at least five editions by 1963. Nevertheless, it represented a past era in a discipline presumably very slow to adapt. Lebon (1963:65) included a section on 'Northern Europeans in the Tropics' in which he argued that whites 'had avoided physical and moral degeneration', although living side by side with blacks, partly by 'refraining from inter-marrying' (p. 68). In another section, Lebon included a table of experimental data on the effect of climate on human activities where we are told, for instance, that in saturated air the maximum temperature before the onset of heat stroke for a 'Nude Subject engaged in heavy manual work in shade' is 86 °F. Fortunately he added the fact that this figure may be a little low, because he had personally observed that 'native populations perform manual work, if not vigorously, at least fairly steadily' above this temperature (p. 49). The point is that what we have here is a very different Geography from that of the present, with some quite different empirical concerns. It has a global focus but it is very much a geographical monologue. Somehow the 'natives' never seem to appear except as instruments of Europeans: the subject remains Europe, the object the world.

INTERLUDE

Emulating the Social Sciences

At approximately the same time as Griffith Taylor and Lebon were peddling the last vestiges of an old global Geography, Schaefer's (1953) article appeared heralding a genuinely new Geography (Johnston, 1979). Schaefer aimed his attack on what he termed the exceptionalism of traditional Geography – the holistic pretensions of regional geography had become a liability in the academic world of specialized social sciences (P. J. Taylor, 1985:102). The subsequent 'quantitative revolution' can be interpreted, therefore, as a modernization of Geography (P. J. Taylor, 1991a). Most commentators consider the key feature of this process to be the adoption of the 'scientific method' as the means for bringing Geography back into the intellectual mainstream. This has been well documented by Johnston (1979) for human geography, and I will say nothing further about it here. Rather I will emphasize a much less discussed feature of the new Geography that is especially pertinent to my argument in this essay: human geographical studies became overwhelmingly state-centric.

The key target of the modernizers was undoubtably the region as the core of geography. With the region out of the way, the spaces that geographers studied could be brought into line with the other social sciences. Since the latter were explicitly creatures of the state, this meant that Geography was finally brought down to the scale of its rivals. For instance, in borrowing models of development from economists, Keeble (1967:247) tells us that 'the key scale group here is the national one'. Hence the domestic space of states became the concern of geographers and the external state dimension, so important for imperialism, was jettisoned. This new, domesticated Geography studied isolated states, converting von Thunen's pragmatic assumption into an unexamined principle of research design.

But it was not just any states that were the subject matter of the new Geography. In a very amusing world map entitled 'The world of spatial science', Derek Gregory (1985:65) locates the new Geography at just five points, three in Germany (von Thunen, Weber and Christaller), one in the USA (Burgess and Hoyt) and one in Sweden (Hagerstrand). This spatial concentration of the underlying theories shows quite explicitly that the new Geography was about 'us', the people of the first world or core of the world-economy, and to a large degree ignored 'them', the vast majority of humanity living beyond the core. Brookfield (1984) describes his experiences of researching in the third world at this time as being those of 'an outside man'. Quite clearly Johnston's (1985) accusation that Geography had disengaged from the world seems borne out.

Of course, the periphery of the world-economy was not completely ignored at this time. Since the theories and models devised from first world locations were deemed to be universal in coverage, it follows that research on the third world could be highly simplified. All that was necessary was to pluck out a tried and tested first world model and impose it on the periphery. And it is in this process that the state-centric nature of the new Geography becomes fully exposed. Most of this research committed what I have termed the 'error of developmentalism' (P. J. Taylor, 1989b). The basic assumption was that *all* states follow the same trajectory of development. Hence by studying social change in a core country, stages of development can be identified that peripheral states will eventually pass through. The most famous such model in social science was Walter Rostow's (1960) 'stages of growth' model; in Geography the most quoted example was based on this – Taaffe, Morrill and Gould's (1963) transport model for 'underdeveloped' countries. In this case it was assumed that Ghana and Nigeria could generate an integrated transport network as found in 'developed countries'. What was missing from their model was the fact that transport development in peripheral states has not been, and will not be for the foreseeable future, an autonomous investment process in which the rest of the world can be ignored. Transport development in the colonies was designed to send commodities to the imperial centre; political independence has not changed the direction of commodity flows and so the role of transport remains the same. Rather than being autonomous networks as studied by the new geographers, they are merely the feeder links into a global transport pattern (P. J. Taylor, 1989b). State-scale studies, especially in the third world, miss an important essence of our modern world. Hence the need to return to the global.

GLOBAL GEOGRAPHY II

Remaking a Discipline at our Fin de Siècle

If we take our commitment to a more contextual history of Geography seriously, then we should expect recent changes in society to be reflected in our studies. This thesis is particularly relevant to our global theme. By the 1970s, global issues are to be found at the top of a wide range of political agendas – this has been called 'the decade of world conferences' (Johnston and Taylor, 1989:1–2). In this context the continuing reproduction of state-centric Geography can be seen as downright embarrassing (P. J. Taylor, 1989b). No wonder that, in assessments of Geography in the 1980s, we can find statements such as Johnston's (1985:328–9) identification of 'the paradox of disengagement' from the world and

Thrift's (1985) plea with British geographers 'to take the rest of the world seriously'. And of course they were beginning to do just that. John Tarrant (1980) and David Grigg (1985) have dealt with the global food problem; M. J. Taylor and Thrift (1982) provided a collection of studies of multinational corporations; and, more generally, Peter Dicken (1986) has looked at the changing international division of labour. Furthermore, Johnston and Taylor (1989) have edited a set of essays all devoted to *A World in Crisis*?. Most indicative of all, perhaps, is the fact that the ultimate 'black sheep' of Geography, geopolitics, has at long last been revived (Hepple, 1986; P. J. Taylor, 1989a). Finally, it can be noted that new textbooks that review the state of the art in the discipline now include their fair share of essays on global questions. Gregory and Walford's (1989) collection of essays matches approximately that of Griffith Taylor's collection referred to previously in proportion of chapters dealing with global concerns. The 1980s were clearly a decade when some geographers, at least, rediscovered their 'geo'. It is at this point that we can be forgiven for thinking that Geography has turned full circle.

The parallels between the two global Geographies are more than a shared interest in a geographical scale. As the global reappears on Geography's agenda, so too does the region. But rather than confirming similarities between past and present, it is here that the differences are clear to see. There is no revival of the implicit environmentalism in the new concern for regions. Johnston (1990) prefers to talk of the geography of regions rather than a new regional geography. There is a concern for places, but not as unique entities. Now the emphasis is upon the making of regions with people as active social agents, not as the playthings of the environment (Anne Gilbert, 1988). Regions are social constructions; they are part of the spatiality of society, not a context in which society is constrained. Regions are going to have an important part to play in the remaking of the discipline at the end of this century, but not as some mystical core of Geography.

Unlike the earlier three subsections of this chapter, where we could describe a global Geography through all its existence, the contemporary remaking of Geography and the role of the global scale in that project has hardly begun. While we can be sure that global studies will be important in the remaking, we cannot be sure what form this will take. Hence any discussion of the new global Geography must in large part be speculation. The disadvantages are obvious, but there is an important advantage for me: I have to a large degree an opportunity to outline what *my* global Geography would look like.

What Could a Global Geography Look Like?

The new global Geography will be a product of what we may call the post-Vasco de Gama era. That is to say that the implicit (and often explicit) racialism in the hierarchical world of imperialism is no longer acceptable as legitimate discourse. After the great revolution of decolonization, we live in a world of formally equal states as certified by membership of the United Nations. This does not mean, of course, that we no longer live in an unequal world, but rather that the processes for reproducing that world have to be more subtle than in the past. Core still dominates periphery, but the mechanisms are more economic than political; formal imperialism has given way to informal imperialism. We live in a very different world from the one in which the first global Geography was constructed – goodbye Vasco de Gama, Halford Mackinder and all.

This new world is distinguished most obviously by the number of interconnections of various forms linking the lives of peoples in all four corners of the world. Although originally set in motion by the worldwide investments of US multinational corporations in the two decades after World War II, for the last two decades it has been a much more complicated affair, as Europe and Japan have come to rival the USA economically. Technological advances in communications have facilitated genuine global decision-making as never before. Although interstate rivalry is most certainly not eliminated, past policies such as the economic autarchy that often accompanied imperialist expansion no longer seem feasible. For some the product is a world of international disorder (Thrift, 1989). For others it is the latest in the stages of capitalism – global capitalism, no less (Graham et al., 1988). While not underestimating the profound changes that have taken place in recent decades, I am going to take a position that affirms the basic continuity of the capitalist world-economy. Today's 'global capitalism' represents a culmination of tendencies that can be traced back several centuries, through which key fundamental characteristics of the system have remained constant. Since these have crucial spatial attributes, I will concentrate upon them in my discussion below. The approach I adopt is Immanuel Wallerstein's (1979, 1983, 1984) world-systems analysis.

The material basis of the world-economy is defined by the myriad commodity chains that criss-cross global space. Each chain consists of a series of production nodes from the point of initial expropriation of nature as raw material to the final point of consumption. At each node the means of production are brought together and, through a specific set of relations of production, value is added to the commodity and passed on down the chain. The social relations at each node determine how much of the additional value added to the commodity is retained at that location and how much seeps through to the next node. Basically the operation of this mechanism can be reduced to two processes, one associated with the core of the world-economy, the other with the periphery. As their names suggest, core-making and periphery-making processes are spatially polarized across the world and have generated two distinctive zones. But between this core and periphery, Wallerstein identifies a semi-periphery where the two processes are relatively balanced. The result is a tripartite hierarchical division of the world into core, semi-periphery and periphery.

We can treat this zonation as a first-order division of the world, within which a second order can be identified in the form of international regions (P. J. Taylor, 1988, 1991b). These are the relatively homogeneous sections of the world zones which encompass distinctive cultural, political and economic characteristics, through which the general zone is continually being reconstructed. We can reproduce this argument with respect to this second-order division to produce a hierarchy of regions. The key point to be made here is that the first-order, tripartite division is a necessary element of the capitalist world-economy (Chase-Dunn, 1989); spatial uneven development is inherent to the system – it is part of its spatiality. Lower-order regions, on the other hand, are contingent: they are historical in the sense that they form and later disintegrate (for example, the Greater Caribbean region that was 'plantation America'); and they may move between zones (for example, Japan). The zones represent the structure of the system; the regions reflect its dynamics.

There is another important, necessary, spatial structure: the Balkanization of the global political space by the inter-state system. This provides the political basis of the system with a formal spatiality that allows for a crucial manoeuvrability for capital in its organization of commodity chains for capital accumulation. Nearly all commodity chains cross state boundaries, which provides for a variety of constraints and opportunities through which the capitalist navigates. Although particular boundaries and their states are the result of historical contingencies, the inter-state system itself is as necessary as the economic spatial uneven development. Between them, these two spatialities define the geography of the system's fundamental political economy.

This is only a very brief glimpse of a new global Geography. On to these bones there have to be added the key institutions through which the two spatialities are reproduced, the households, classes and peoples (nations, ethnic groups), and the various political and cultural movements through which they are mobilized (P. J. Taylor, 1991c). My characterization of this world through just two spatialities must not be read as defining a simple world of political economy. We may be living in 'one world', but it is a highly complex one. Furthermore, we can only begin to appreciate this complexity through understanding regions at various levels that between them make up the world-economy. The new global Geography should describe a highly connected, hierarchical and ever-changing spatial mosaic

Geographical Dialogue in a World Subject

Whatever new global Geography transpires, it must transcend the Vasco de Gama era: a geographical monologue will no longer be acceptable. My global Geography explicitly replaces Eurocentric domination with a liberation discourse that converts the monologue into a geographical dialogue.

I have derived the geographical dialogue concept from E. H. Carr's classic (1962) discussion *What is History*?. He describes History as a continual dialogue between the present and the past. The subtlety of his argument is that the present is always *in* the past as described by historians, who inevitably project their contemporary concerns and interests on to the particular pasts they describe. Hence there can be no ultimate History when a period is finally, completely known; rather every generation of historians adds its concerns to the History in a never-ending process of evolution and revolution. In this way, History remains a dynamic rather than a static discipline; it is alive, not dead.

By analogy, Geography's dialogue is between places. It is a dialogue between the location where the geographer has been socialized and where he or she lives, and the place of interest. The home place provides interests and concerns that colour the agenda according to which the subject place is studied. The stages in developmentalist models imposed on third world countries illustrate this process well – how long has India been 'taking off' now? To extend the analogy, there will be many Geographies of a place depending on where it is viewed from. The key point, therefore, is that every place is many places: there is no one objective description, just alternative Geographies. Like History, Geography is dynamic and therefore alive.

In an unequal world such as the one we live in, geographical dialogue can easily degenerate into a monologue reflecting the power relativities of places. That was what happened to the Geography of imperialism. As we have seen, Orientalism represents a sort of limiting case in the projection of home-place concerns on to the subject place: 'that Orientalism makes sense at all depends more on the West than on the Orient' (Said, 1978: 22). And, of course, such projection continues to inform our knowledge of Islam despite the end of the Vasco de Gama era (Said, 1981). But this is where our new global Geography comes into the picture. Whether our particular global Geography derives from a social theory of global mutual interests, such as that underlying the Brandt Report, or the oppositionalist theory of world-systems analysis, empathy between places must be at the heart of our dialogues. In this way, and despite its imperialist past, Geography can become a liberating discipline for teaching an equal dialogue between places. Differences, similarities and connections between places can be studied to promote understanding without generating feelings of superiority, whatever the levels of material differences (P. J. Taylor, 1989c). Our recent experiences of first world triumphalism in the Gulf War suggest this is an urgent educational task for geographers (Mitchell and Smith, 1991). Only in this way can Geography attain the status of the genuine world subject it has always had the potential to become but which it has never achieved.

NOTES

1 Throughout this chapter, Geography with a capital G refers to the academic discipline.

REFERENCES

Alexander, R. J. 1973: *Aprismo: the ideas and doctrines of Victor Raul Haya de la Torre*. Kent, OH: Kent State University Press.

Berdoulay, V. 1980: The contextual approach. In D. R. Stoddart (ed.), *Geography, Ideology and Social Concern*, Oxford: Basil Blackwell, 8–16.

Bergman, E. F. 1975: *Modern Political Geography*. Dubuque: Wm Brown.

Bissio, B. (ed.) 1988: *Third World Guide*. Montevideo: Third World Editors.

Brookfield, H. C. 1984: Experiences of an outside man. In M. Billinge, D. Gregory and R. Martin (eds), *Recollections of a Revolution*, London: Macmillan, 27–38.

Bunge, W. 1965: Racialism in geography. *Crisis*, 2, reproduced in 1984 in *Contemporary Issues in Geography and Education*, 1 10–11.

Carr, E. H. 1962: *What is History?* London: Penguin.

Chase-Dunn, C. 1989: *Global Formation: structures of the world-economy*. Oxford: Basil Blackwell.

Claval, P. 1984: The coherence of political geography. In P. J. Taylor and J. House (eds), *Political Geography: recent advances and future directions*, London: Croom Helm, 8–24.

Dicken, P. 1986: *Global Shift: industrial change in a turbulent world.* London: Paul Chapman.

Gilbert, Alan 1987: Review of *World in Crisis?*. *Environment and Planning A*, 19, 130–2.

Gilbert, Anne 1988: The new regional geography in English- and French-speaking countries. *Progress in Human Geography*, 12, 208–28.

Graham, J., Gibson, K., Horvath, R. and Shakow, D. M. 1988: Restructuring in US manufacturing: the decline of monopoly capitalism. *Annals of the Association of American Geographers*, 78, 473–90.

Grano, O. 1980: External influence and internal change in the development of geography. In D. R. Stoddart (ed.), *Geography, Ideology and Social Concern*, Oxford: Basil Blackwell, 17–36.

Gregory, D. 1985: People, places and practices: the future of human geography. In R. King (ed.), *Geographical Futures*, Sheffield: The Geographical Association, 56–76.

Gregory, G. and Walford, R. (eds) 1989: *Horizons in Human Geography*. London: Macmillan.

Grigg, D. 1985: *The World Food Problem, 1950–1980.* Oxford: Basil Blackwell.

Hepple, L. W. 1986: The revival of geopolitics. *Political Geography Quarterly*, 5 (supplement), 21–36.

Herbertson, A. J. 1910: Geography and some of its present needs. *Geographical Journal*, 36, 468–79.

Hudson, B. 1977: The new geography and the new imperialism, 1870–1918. *Antipode*, 9, 12–19.

Jones, S. B. 1954: A unified field theory of political geography. *Annals of the Association of American Geographies*, 44, 111–23.

Johnston, R. J. 1979: *Geography and Geographers*. London: Edward Arnold.

Johnston, R. J. 1985: To the ends of the earth. In R. J. Johnston (ed.), *The Future of Geography*, London: Methuen, 326–38.

Johnston, R. J. 1990: The challenge for regional geography: some proposals for research frontiers. In R. J. Johnston, J. Hauer and G. A. Hoekveld (eds), *Regional Geography: Current developments and future prospects*, London: Routledge, 122–39.

Johnston, R. J. and Taylor, P. J. (eds) 1989: *A World in Crisis?: geographical perspectives* (second edition). Oxford: Basil Blackwell.

Kabbani, R. 1988: *Europe's Myth of Orient*. London: Pandora.

Keeble. D. E. 1967: Models of economic development. In R. J. Chorley and P. Haggett (eds), *Models in Geography*, London: Methuen, 243–300.

Lebon, J. H. G. 1963: *An Introduction to Human Geography* (fifth edition). London: Hutchinson.

Mitchell, D. and Smith, N. 1991: Bush League political geography. *Political Geography Quarterly*, 10, 338–41.

Parker, W. H. 1982: *Mackinder: Geography as an aid to statecraft.* Oxford: Oxford University Press.

Pelzer, K. J. 1951: Geography and the Tropics. In G. Taylor (ed.), *Geography in the Twentieth Century*, London: Methuen.

Rostow, W. W. 1960: *The Stages of Economic Growth*. Cambridge: Cambridge University Press.

Sachs, I. 1976: *The Discovery of the Third World*. Cambridge, MA: The MIT Press.

Said, E. W. 1978. *Orientalism*. New York: Pantheon Books.

Said, E. W. 1981: *Covering Islam*. London: Routledge & Kegan Paul.

Schaefer, F. K. 1953: Exceptionalism in geography: a methodological examination. *Annals of the Association of American Geographers*, 43, 226–49.
Stoddart, D. R. 1982: Geography – a European science. *Geography*. 67, 289–96.
Taaffe, E. J., Morrill, R. L. and Gould, P. R. 1963: Transport expansion in underdeveloped countries: a comparative analysis. *Geographical Review*, 53, 503–29.
Tarrant, J. R. 1980: The geography of food aid. *Transactions, Institute of British Geographers*, NS 5, 125–40.
Taylor, G. (ed.) 1951: *Geography in the Twentieth Century*. London: Methuen.
Taylor, M. J. and Thrift, N. J. (eds) 1982: *The Geography of Multinationals*. London: Croom Helm.
Taylor, P. J. 1985: The value of a geographical perspective. In R. J. Johnston (ed.), *The Future of Geography*, London: Methuen, 92–110.
Taylor, P. J. 1988: World-systems analysis and regional geography. *Professional Geographer*, 40, 259–65.
Taylor, P. J. 1989a: *Political Geography: world-economy nation-state and locality* (second edition). London: Longman.
Taylor, P. J. 1989b: The error of developmentalism in human geography. In D. Gregory and R. Walford (eds), *Horizons in Human Geography*, London: Macmillan, 303–19.
Taylor, P. J. 1989c: Geographical dialogue. *Political Geography Quarterly*, 8, 103–5.
Taylor, P. J. 1991a: A future for Geography. *Terra*, 103, 21–31.
Taylor, P. J. 1991b: A theory and practice of regions: the case of Europe. *Environment and Planning D: Society and Space*, 9, 183–96.
Taylor, P. J. 1991c: Understanding global inequalities: a world-systems approach. *Geography*, 76, 10–21.
Thrift, N. J. 1985: Taking the rest of the world seriously? the state of British urban and regional research in a time of economic crisis. *Environment and Planning A*, 17, 7–24.
Thrift, N. J. 1989: The geography of international economic discorder. In R. J. Johnston and P. J. Taylor (eds), *A World in Crisis?: geographical perspectives*, Oxford: Basil Blackwell, chapter 2.
Turner, F. 1980: *Beyond Geography*. New York: Viking.
Unstead, J. F. and Taylor, E. G. R. 1910: *General and Regional Geography*. London: George Philip.
Vujakovic, P. 1989: Mapping for world development. *Geography*, 74, 97–106.
Wallerstein, I. 1979: *The Capitalist World-economy*. Cambridge: Cambridge University Press.
Wallerstein, I. 1983: *Historical Capitalism*. London: Verso.
Wallerstein, I. 1984: *The Politics of the World Economy: the states, the movements and the civilisations*. Cambridge: Cambridge University Press.
Whittlesey, D. 1939: *The Earth and the State*. New York: Henry Holt.
Wolf, E. 1982: *Europe and the Peoples without History*. Berkeley, CA: University of California Press.

10

Changing Ourselves: A Geography of Position

Peter Jackson

There is an ecological problem, an urban problem, an international trade problem, and yet we seem incapable of saying anything of depth or profundity about any of them. When we do say something, it appears trite and rather ludicrous. In short, our paradigm is not coping well. It is ripe for overthrow.

Harvey, 1973:129

When David Harvey wrote these prescient words, nearly twenty years ago, there was an undoubted sense of optimism underlying his call for a revolution in geographic thought. If geographers changed their paradigm, Harvey implied, they could change the world. Today, the mood is far less optimistic. Even the sub-title of this collection ('a changing world: a changing discipline?') implies a reversal of Harvey's logic. Rather than changing the discipline in order to change the world, geographers now cast themselves in a much less active role, passively responding to changing circumstances. What kinds of change in the wider world account for such a dramatic shift in academic self-confidence? Why do we no longer feel so sure of our ability to control the changing environment? Environmental catastrophes from Three Mile Island to Chernobyl, from Love Canal to Bhopal, have certainly played a significant role in undermining public confidence in the omniscience of science, casting doubt on the plausibility of experts and weakening our faith in the onward march of progress. Even the idea of 'controlling' the environment is now regarded with scepticism, implying an uncritical adherence to an excessively technocratic form of scientific rationality (Gregory, 1980). Calls for greater 'environmental humility' (Relph, 1981), for the integrity of various forms of 'local knowledge' (Geertz, 1983) and for an indigenous 'people's science' (Richards, 1985) have become increasingly common. Far from our controlling the environment, its very sustainability is now widely called into question (Redclift, 1987). If, as I intend to argue here, these environmental concerns are part of a wider crisis of confidence affecting western scientific thought at the end of the twentieth century, then our own disciplinary crisis can best be understood by placing it in this wider intellectual and political context.

Some of the earliest and most fundamental challenges to the authority of science have come from feminist sources (such as Merchant, 1982; Harding, 1986). More recently, under the banner of postmodernism, anthropologists have begun to question their individual and collective 'ethnographic authority' (Clifford, 1983), admitting that their representations of other cultures may actually provide a better guide to the anthropologist's own society's cultural preoccupations and domestic concerns than a faithful portrait of the 'exotic' cultures they claim to represent. This insight was originally made by Edward Said in his seminal work on Orientalism (Said, 1978). It has since been taken up by geographers such as Kay Anderson, who shows how much the contemporary social geography of Vancouver's Chinatown owes to the history of western ideas about the Orient (Anderson, 1991).

Anthropologists now speak of a 'crisis of representation' throughout the human sciences (Marcus and Fischer, 1986), a challenge to which geographers are belatedly responding (Barnes and Duncan, 1992). Their attempts to chart *The Condition of Postmodernity* (Harvey, 1989) and to write avowedly *Postmodern Geographies* (Soja, 1989) have attracted considerable attention outside the discipline – some of it flattering (such as Terry Eagleton's dust-jacket assessment of David Harvey's 'devastating' book); some considerably more reserved (such as the art historian Rosalyn Deutsche's, 1990, searing commentary on 'Men in space'). Against this background, it is no surprise that geography and geographers have been experiencing their own disciplinary crisis. Within the last decade, geography has been rethought, remodelled, reconstructed and remade; old (positivist) models have been challenged by new ones (from political economy), and new horizons have been charted in physical and human geography.[1] In geography, as elsewhere, there is a sense of intellectual fragmentation, of disciplinary ferment, as 'All that is solid melts into air' (Berman, 1982).

In this brief chapter I can do no more than highlight a number of the challenges that have been directed towards some of the concepts that geographers have traditionally held dear – like landscape and nature

– and suggest some possible ways out of the current labyrinth, considering the merits of an alternative geographical vocabulary, where words like 'margin' and 'position' take on a new significance. Above all, I will argue, it is in the recognition of how we are ourselves 'placed' in terms of the ever-changing social relations of power that we may find a viable way of responding to the dilemmas with which we are currently faced. Identifying the problem in terms of *positionalities* requires us to be prepared to change ourselves before we set about changing our discipline or the wider world. For inspiration, I will be drawing on a variety of feminist sources while recognizing that we cannot simply plunder these, tacking them on to existing ways of thinking. Taken seriously, they have the potential to transform the very nature of what we study and our way of being in the world.

CRISIS AND RECONSTRUCTION

If, as David Harvey suggests, the 1970s witnessed an urban crisis, an energy crisis, an environmental crisis, then by the 1980s the whole world appeared to be in crisis (Johnston and Taylor, 1986). Intellectually, the crisis was experienced as a series of attacks on received wisdom, as old certainties began to crumble, and what once seemed absolute gave way to a host of relativities. Within geography, a succession of challenges was launched on some of the discipline's core concepts. Here I will focus on landscape and nature, but a similar exercise could be undertaken for other key concepts, such as place, space, region or locality.

One of the most effective challenges was posed by Denis Cosgrove's critique of the idea of landscape. Cosgrove's writings represent a sustained attempt to deconstruct the understanding of landscape that characterized an earlier generation of cultural and historical geographers such as Carl Sauer, Clifford Darby and Andrew H. Clark. Cosgrove rejected the simple distinction which Sauer drew between 'natural' and 'cultural' landscapes, showing that *every landscape is socially constructed*. Taking the example of landscape painting, he showed how the development of perspective in the fifteenth century coincided with the rise of merchant capitalism and the need for accurate measurement of land-as-property (Cosgrove, 1985a). Rather than tracing a uni-directional 'human impact on the landscape' (a tradition which he and others have parodied as the 'landscape with figures' school), Cosgrove showed how landscape and social formation were dialectically related (Cosgrove, 1985b). More recently (Cosgrove, 1990), he has argued that in the transition from modernity to postmodernity biological and cybernetic models of environmental and spatial organization (organism, system) have given way to metaphors derived from literature and the arts (spectacle, theatre, text).

While Cosgrove has challenged the 'naturalness' of landscape, others have begun to question the 'naturalness' of nature itself. Neil Smith (1984) began the current round of critique by adding his own materialist reading to the long tradition of Marxist thought about the social production of nature. But the debate has been given renewed vigour by the recent work of Margaret FitzSimmons (1989a, 1989b). While Smith develops his ideas about socially produced nature through a reworking of Marx's theory of uneven development, FitzSimmons starts out from Raymond Williams's essays on 'The idea of nature' (1980), extending his work via a variety of feminist sources. She explores the parallel between man's subordination of nature and his subordination of women, a material context from which all kinds of ideologies have been elaborated: Mother Nature, virgin land, Earth Mother, and many more. In similar fashion, feminist historians like Annette Kolodny have undertaken new interpretations of the American frontier experience, showing how women's experience was radically different from that of men (Kolodny, 1984). Geographers have also begun to dissect the gendered nature of environmental experience, with studies like *The Desert is No Lady* (Norwood and Monk, 1987) paving the way for a host of new interpretations, some involving new methodologies (such as Burgess, Limb and Harrison, 1988a, 1988b).

These are just two examples of traditional geographical concepts (landscape and nature) which have been radically deconstructed, drawing on the insights of Marxist and feminist scholarship. But each of these examples does much more than merely challenge existing categories. Rather than simply showing 'a discipline in crisis', they reveal the possibility of new kinds of work, as reconstruction follows critique. For example, FitzSimmons couches her argument in terms of the need for a reconciliation between 'human' and 'physical' geography, arguing that the recent debate about 'society and space' should be followed by a similar debate about 'society and nature', while Cosgrove's work has led to new work in the field of landscape symbolism and iconography (such as Cosgrove and Daniels, 1988).

These studies make my point about the positive potential of critique and reconstruction, working largely within existing disciplinary boundaries. But to advance my argument about the politics of position requires that we be prepared to transcend the boundaries of what has traditionally constituted the subject matter of geography. I can make my argument most effectively in relation to the social construction of 'race' and gender, both categories whose 'naturalization' has been challenged by a range of feminist and post-colonial writers. To explore their relevance for a reconstituted human geography involves the adoption of a new (but equally spatialized) vocabulary concerned with centres and margins, boundaries and positions. But I start with a consideration of perspectives, standpoints and 'ways of seeing'.

PERSPECTIVES, STANDPOINTS AND WAYS OF SEEING

John Berger's exploration of our conventional 'ways of seeing' (1972) has long been of interest to geographers. Denis Cosgrove cites the work in his discussion of the evolution of the landscape idea (Cosgrove, 1985a); Hugh Prince refers approvingly to Berger in his discussion of art and agrarian change in eighteenth-century England (Prince, 1988); and the phrase recurs in several contributions to a recent collection of essays on cultural geography (Anderson and Gale, in press). But there is a growing feminist literature on our acquired 'ways of seeing' that these authors largely ignore. While some of these contributions have been made in specific contexts, such as feminist art history (for example, Pollock, 1988), others apply more generally to methodological debate within the social sciences.

An example is Sandra Harding's argument for a feminist standpoint in her book on *Feminism and Methodology* (1987), where she challenges the notion that feminist research is fatally flawed by an inherent relativism. She shows how, historically, relativism has only been recognized as a problem by dominant groups when the universality and hegemony of their particular views are being challenged. Just as cultural relativism emerged within anthropology as a response to the colonial encounter (Asad, 1973), so has masculinist social science responded defensively to the rise of feminism. In this case, Harding argues, relativism is a sexist response, attempting to preserve the legitimacy of androcentric concerns in the face of contrary evidence (1987:10).

Again, this is not merely a negative argument, as Harding goes on to assert the superiority of a feminist understanding in its approach to certain problems. Drawing on Nancy Hartsock's (1983) work, Harding makes the crucial distinction between a standpoint and a perspective: 'a feminist standpoint is not something anyone can have by claiming it, but an achievement' (1987:185). Unlike a perspective, it has to be struggled for; to be earned. And, at least in certain cases, there are good grounds for believing that women's experience, informed by a feminist standpoint, offers a superior mode of understanding to the 'abstract masculinity' of gender-blind social science. Masculinist vision, Harding argues (1987: 171), is both partial and perverse, valuing production over reproduction, the public over the private, the arena of work over the domain of home, and so on. Hartsock's argument lends support to those who maintain that research which proceeds from an explicit political commitment (in this case to feminism) may produce sounder results than apparently disinterested research which fails to examine its own partialities. This is what is implied in the development of a politics of position.

CENTRES AND MARGINS: TRANSGRESSING BOUNDARIES

One of the most compelling arguments for a politics of position comes from the work of bell hooks.[2] In a particularly 'geographical' essay, she explains the difference between being forced into the margins by the oppressive structures of a white, capitalist, patriarchy and 'choosing the margin as a space of radical openness' (hooks, 1990). This, she explains, involves a process of transgression, deliberately moving out of one's place:

> For many of us, that movement requires pushing against oppressive boundaries set by race, sex, and class domination. Initially, then, it is a defiant political gesture. Moving, we confront the realities of choice and location. Within complex and ever shifting realms of power relations, do we position ourselves on the side of the colonizing mentality? Or do we continue to stand in political resistance with the oppressed, ready to offer our ways of seeing and theorizing, of making culture, towards that revolutionary effort which seeks to create space where there is unlimited access to the pleasure and power of knowing, where transformation is possible? (hooks, 1990:145).

The language of transgression has become a leitmotif of postmodernism in works like Stallybrass and White's *The Politics and Poetics of Transgression* (1986). But, while hooks freely admits that language is a place of struggle – that 'finding a voice' is a besetting problem for those whose discourse has been systematically marginalized – she is clearly impatient with those who argue that politics is nothing more than a struggle over language. She criticizes recent developments in ethnography and cultural studies, for example, for their complicity in reinscribing patterns of colonial domination, where 'the "Other" is always made object, appropriated, interpreted, taken over by those in power, by those who dominate' (hooks, 1990:125). While cultural studies has rightly rejected essentialist notions of difference, she argues, this critique should not become an excuse for ignoring difference or for dismissing the authority of experience.

Her argument applies *a fortiori* to the academic literature on postmodernism, which speaks about the radical possibilities of a 'politics of difference' while remaining, as a discursive practice, clearly dominated by the voices of white, male intellectuals who speak to one another in terms of coded familiarity (hooks, 1990:24). Rather than simply rejecting postmodernism as 'the latest hip racism' (p. 133), however, hooks adopts a more conciliatory tone. She welcomes the opportunity postmodernism offers for incorporating the voices of the displaced, marginalized, exploited and oppressed. But she warns against the kind of postmodernist critique of the 'subject' that calls on marginalized groups to renounce their sense of identity as an untenable form of essentialism at the very time when they are 'coming to voice' for the first time (p. 28). There is, she argues, a crucial difference between the repudiation of essentialism and the recognition that in certain circumstances people's experience affords them a privileged critical location from which to speak.

There may, then, be grounds for preferring one standpoint over another; for privileging marginalized discourses over those which have traditionally commanded centre stage. Failure to grasp this point seems to me the principal weakness in Steven Connor's argument about centre and margin in his study of *Postmodernist Culture* (1989). He argues that while postmodernism celebrates the return to the centre of formerly marginalized discourses, there is nothing intrinsically superior about such marginality. He may be right if he means that voices should not be privileged simply because of where they originate, irrespective of their content. But he is surely wrong if he implies that we should not be sensitive to the contexts from which particular discourses emerge. African American politicians like Jesse Jackson or Louis Farrakhan *may* be guilty of anti-Semitism, for example. But we will never understand what they are trying to say if we ignore the racism against which their own discourses have emerged.

These are admittedly murky waters and there is a danger of seeming to apply different standards to arguments from different quarters. It is difficult for me as a white man, for example, to criticize the implicit racism of William Julius Wilson's argument in *The Truly Disadvantaged* (1987). Writing as an African-American, his arguments about the condition of the black underclass in American cities have an apparent authority that cannot be ignored. To criticize his argument without considering the circumstances in which it was made, including the author's experience as an African-American, would clearly be reprehensible. But so would the application of a double standard, which I explicitly reject (Jackson, 1989a:32–3).

A second example of this kind of dilemma concerns the question of whether men can be feminist or whether they are inevitably excluded from more active participation in the women's movement. Sandra Harding (1987:11) argues that men can make important contributions to feminist research and scholarship. She rejects as essentialist the idea that one's ability and willingness to contribute to feminist understanding are sex-linked traits. She is prepared to extend the designation 'feminist' to men if they satisfy whatever standards women must satisfy to earn the label. She admits, though, that women would be wise to look especially critically at 'feminist' analyses produced by members of the oppressor group. While I thoroughly endorse the need for scepticism in such circumstances, my own position is somewhat different. I would deny men's right to the designation 'feminist', arguing that they should support feminist demands for equal rights but respect the autonomy of feminism as a women's movement (Jackson, 1991).

What, then, are the implications of these arguments for the achievement of a more truly human geography? First, we should not deny or marginalize the history of feminist scholarship within the discipline. Feminist geographers began their assault on the 'naturalization' of gender categories some years ago (Women and Geography Study Group Institute of British Geographers, 1984), demonstrating that the distinctions between home and work, public and private, productive and reproductive labour are culturally, historically and geographically specific, not the result of some immutable natural law. Similarly, the idea of a number of discrete, naturally occurring human 'races' has been challenged by theories that insist on the social construction of 'race' (such as Jackson, 1987).

Both feminist and anti-racist literatures pose sharp questions about the role of professional social science: how can professions like geography, which are so heavily dominated by white, middle-class, able-bodied men and whose histories are so deeply implicated in the politics of empire, begin to address these issues in a critical manner? Here, we need to look outside our disciplinary boundaries, drawing inspiration from new sources, including the work of feminist and post-colonial writers.

A second implication of my argument would be to resist the tendency to construct firm boundaries around our disciplinary discourse. There is much to gain from the transgression of boundaries and much to fear from those who seek to police them. But we should also focus on how we are ourselves placed with respect to certain crucial forms of power and inequality. What 'cultural capital' do we have as professional geographers, and how do we exercise that power? I have begun to address these issues in my own work, 'writing as a white, middle-class man, working in a privileged though increasingly beleaguered profession and living in the capital of what was once the heart of Empire' (Jackson, 1989a:x). In my own research I have tried to come to terms with how I am positioned by shifting the focus away from the investigation of 'ethnic minority' distributions and towards an analysis of the social and spatial constitution of British racism (Jackson, 1987, 1992). I have also begun to examine the scope for challenging racism and sexism through my teaching of geography (Jackson, 1989b). Similarly, as a man, my priorities for research on gender relations have been to explore the spatial basis of dominant forms of masculinity rather than the geographies of women's consciousness and experience.

Besides these moral imperatives, confronting questions of gender and 'race' oppression provides an avenue for understanding how our society works, how it has developed as it has, and how it may be changed. But even here, the politics of position are more complex than they seem. As geographers and other social scientists have begun to relinquish their traditional role as the neutral arbiters of contemporary 'race' or gender relations, they run the risk of becoming equally uncritical champions of subcultural resistance (Bourne and Sivanandan, 1980). An example will illustrate the complexities to which I allude.

THE GHOSTS OF HANDSWORTH

In 1986, immediately following the 'riots' that took place throughout mainland Britain, John Akomfrah directed a film for the London-based Black Audio Film Collective entitled *Handsworth Songs*. Critical reaction to the film was mixed and nicely illustrates the problems involved in articulating an effective politics of position. The film's narrator repeatedly asserts that 'There are no stories in the riots, only the ghosts of other stories'. Using a variety of visual images (including footage from previous documentaries, newsreels and freshly assembled 'still lives'), together with an innovative use of music and poetry, the film aims to give an alternative perspective on the riots from the conventional voices of police chiefs, politicians and news media. Despite these laudable ambitions, reactions to the film (as to the events themselves) were highly charged.

Salman Rushdie argued that John Akomfrah had failed to develop an appropriate 'language' in which to tell the untold stories of the riots: 'It just isn't enough to be black and blue, or even black and angry . . . [If] you want to tell the untold stories, if you want to give voice to the voiceless, you've got to find a language' (*Guardian*, 12 January 1987). Rushdie quotes from Black Audio's press release which states that 'The film attempts to excavate hidden ruptures/agonies of "Race" ', looking at the riots as 'a political field coloured by the trajectories of industrial decline and structural crisis' and 'repositioning the convergence of "Race" and "Criminality" '. In place of the 'dead language of race industry professionals' with its tortuous language of 'ruptures' and 'trajectories', Rushdie would have preferred to hear more of the rich language of the film's subjects themselves. He admits that it is not easy for black voices to be heard in Britain but argues that we should resist the desire to cheer 'whenever somebody says what we all know, even if they say it clumsily and in jargon'. He fears that such uncritical celebration, however well-intentioned, will only make Britain's black communities lazy.

Writing to the *Guardian* letters page (19 January 1987), Darcus Howe, editor of *Race Today* and a member of the professional 'race industry' alluded to by Salman Rushdie, agreed with the controversial novelist. Praising Rushdie's 'useful and timely intervention', Howe thought that black film-makers could only benefit from Rushdie's intervention, which functioned both as a critique and as a foundation for a new critical tradition. An alternative position was articulated by Stuart Hall in an earlier letter (15 January 1987) where he criticized Rushdie's attack on Black Audio Film Collective, praising the struggle that the film represents to break with the tired style of riot-documentary. Hall applauded the film-makers' efforts to develop a new language in which the black experience is told as an *English* experience. He cites the way documentary footage has been retimed, retinted and overprinted so as to formalize and distance it. He praises the narrative interruptions; the originality and unpredictability of the soundtrack; the attempt to 'give voice' to new subjects; and the intercutting with the 'ghosts' of other stories.

The critical controversy over *Handsworth Songs* demonstrates a number of points. First, it exemplifies the growing diversity of voices within Britain's black communities. Second, it reveals a range of opinions over the need to develop new 'languages' in which to convey new ideas (and some disagreement over the film's success in doing this). Thirdly, it shows that many positions can be adopted with respect to any single issue even among those with a common experience and a shared commitment to challenging racism. But notice, too, how the authors position themselves and how they are positioned by each other. Hall criticizes Rushdie's comfortable elitism, speaking from a 'well-deserved but secure position in the literary firmament'. Howe replies by drawing attention to Hall's 'equally well-deserved but secure position in the academic firmament', identifying himself as 'an activist in the black movement for over 20 years, organising and developing political, cultural and artistic thrusts which have emerged from within our black communities'. It is a relatively insignificant example of the politics of position, but one from which there is much to learn.

POLITICS AND METHOD

In order to address these issues, geographers have begun to draw on a variety of literary and cultural theories, bringing to them our own disciplinary concerns with space and place, distance and territory, centres and margins. Among these theories I would emphasize historical and cultural studies of the social construction of 'racial' difference (Gates, 1986), literary studies of textual strategy and narratives of nation (Bhabha, 1990), and the rapidly developing anti-imperialist field of 'subaltern studies' (Spivak, 1987). The post-colonial context of 'subaltern studies' is of particular interest, given our discipline's profoundly imperialist past (Driver, 1992). Spivak's choice of terminology is deliberate, referring to the oppression of 'subaltern' groups in order to emphasize specific histories of military conquest and imperial domination. She asks how the subaltern can speak, in the knowledge that her voice will immediately be appropriated by her oppressors, including first world intellectuals interested in 'capturing' the voice of the colonial Other.

Spivak's work raises profound epistemological problems for those who seek to articulate a politics of position. How can we, who have a professional interest in 'representing the Other', carry out our work without imposing a form of 'epistemic violence' on the 'subjects' of our research (Spivak, 1988:283)? Anthropologists have responded to these representational problems by experimenting with different 'textual strategies', seeking alternatives to the implied authority of a single ethnographic voice by adopting more complex forms of narrative and making room for a multiplicity of voices (for example, Clifford and Marcus, 1986). But these experiments offer only a partial solution to the wider questions raised by Spivak's work.[3] They run the risk of what Bryan Palmer (1990) describes as a 'descent into discourse', where an obsession with textual strategy and the poetics of geographical description reduces life to language, obscuring wider relations of power, exploitation and inequality.

As with the insights of feminist research, it is not simply a matter of bolting on the findings and approaches of the latest intellectual fad. If new voices are to be heard, the discipline itself must be reshaped, its boundaries transcended and redrawn. As feminist research has shifted the boundaries between public and private, home and work, sex and gender, so too will the empowerment of other marginalized discourses involve a fundamental rethinking in our attitudes to 'race' and empire, territory and nation.

Greater self-consciousness about narrative style and textual strategy is beginning to form part of the geographical agenda (cf. Gregory, 1989; Sayer, 1989). There are even some signs of what a progressive post-colonial geography might look like. Jonathan Crush (1991), for example, has shown how debates among geographers in South Africa relate to wider debates about the nature of the South African state. Beyond the question of textual strategy, Crush is interested in how far South Africa's 'critical moment' has helped fashion the academic positioning and intellectual choices of progressive geographers. Crush shows that the relations between text and context are dialectical, as authors seek extra-textual support for their arguments. Likewise, much of the recent postmodern literature in geography and neighbouring disciplines has sought to demonstrate its critical edge by a strategic use of political metaphors. How often these days do we read about 'struggle' and 'resistance', where the politics involved is confined to the page?

The French sociologist Pierre Bourdieu explores this theme in *Homo Academicus* (1988), exposing the 'tribal secrets' of the intellectual life and the forms of power that are wielded in the name of science. Bourdieu reveals how social scientists have attempted to validate their work by seeking engagement with a world beyond the ivory tower, even when that association is no more than literary. He is particularly interested in the events of May 1968 and their relation to the internal politics of French academic life. Inverting the usual relationship between academic position and political commitment, he claims 'it is not, as is usually thought, political stances which determine people's stances on things academic; but their positions in the academic field which inform the stances that they adopt on political issues in general as well as on academic problems' (Bourdieu, 1988:xvii–xviii). The point may be worth pondering in relation to our own position within contemporary geography: what 'cultural capital' do we possess and what are our commitments beyond the confines of our professional lives?

CONCLUSION

Let me return, in conclusion, to David Harvey, whose latest work seems to me to address some of the same questions that I have been exploring here, though using rather different sources. Harvey argues that responding to global, environmental and social issues requires fundamental changes, personal commitment, and the kind of moral courage that is not compatible with much of our learned ways of seeing and doing. We cannot, for example, simply recycle a small proportion of our domestic waste and feel that we have done our bit for the environment. Harvey reminds us that the contradictions of 'environmentalism' run much deeper:

> We cannot reasonably go to church on Sunday, donate copiously to a fund to help the poor in the parish, and then walk obliviously into the market to buy grapes grown under conditions of apartheid. We cannot reasonably argue for high environmental quality in the neighborhood while still insisting on living at a level which necessarily implies polluting the air somewhere else. (Harvey, 1990:423)

Yet this is exactly the sort of contradiction that we, as academics, seem prepared to tolerate. We accept the validity of certain kinds of analysis (on the iniquities of capitalism, patriarchy or racism, for example) and yet refuse to accept their consequences for our own lives.

A politics of position, as outlined here, might help us to address these issues more effectively. Stuart Hall reaches a similarly up-beat conclusion in his reading of the politics of postmodernism:

> Far from there being no resistance to the system, there has been a proliferation of new points of antagonism, new social movements of resistance organised around them – and, consequently, a generalisation of 'politics' to spheres which hitherto the Left assumed to be apolitical: a politics of the family, of health, of food, of sexuality, of the body. (Hall, 1989:130)

Contemporary society is characterized by multiple forms of exploitation and oppression, which suggest that our politics should also be increasingly positional. But what we lack, Hall argues, is any *overall map* of how these power relations connect and of their resistances.[4] His choice of metaphor has been questioned by those who take a more pessimistic line on the politics of postmodernism. For Steven Connor, for example, the very terms space and territory, centre and margin, inside and outside, position and boundary 'conjure up an oddly antique-seeming map of the world' (Connor, 1989:227). From this perspective, a map is an inappropriate metaphor in so far as it implies a position outside of, or suspended above, the field that is being surveyed. Connor concludes that the problem for postmodern theory is to construct a map of the world *from inside that world*.

There is an interesting parallel here with feminists like Sandra Harding (1987), who argue that placing ourselves on the same analytical plane as those we claim to represent and whose experience we seek to interpret is ethically and methodologically axiomatic. In other respects, the encounter between feminism and postmodernism has been far from easy. Linda Nicholson neatly summarizes the problem: if postmodernism is 'the view from eveywhere', feminism implies a very particular 'politics of location' (Nicholson, 1990:8). Both movements challenge us to consider from where we speak and whose voices are sanctioned. Yet feminism's critical commitment runs counter to the political paralysis associated with some brands of postmodern theory. If they have little else in common, however, feminism and postmodernism both teach us that the arena of political struggle is broader than we once thought, encompassing constructions of gender, 'race' and class, and urging the investigation of their variable intersections in particular places at particular times.

To write the geographies of the 1990s, we must face up to the challenge of constructing new maps of the connections between local resistances and individual sites of struggle, welcoming the recognition of 'difference' without surrendering to a politics of indifference. Those of us who wish to change the discipline and have ambitions to change the world should start from modest beginnings, recognizing our own *positionality* with respect to the fundamental inequalities of gender, 'race' and class. For if those dimensions are a source of power to us, they are as surely a source of oppression to those around us. If our geographical imagination is to develop in ways that are genuinely and constructively oppositional (Cocks, 1989), we should begin by changing ourselves.

ACKNOWLEDGEMENTS

Many thanks to everyone who contributed to the discussion of this paper at the IBG conference in Sheffield and to those who read earlier drafts of this chapter, especially Linda McDowell and Sarah Whatmore.

NOTES

1 For references to this literature, see chapter 1 in the present volume.
2 bell hooks is the pseudonym of the African-American feminist Gloria Watkins, who chooses to write under her grandmother's name in the lower case. This allows her to engage in some novel textual strategies, including a series of 'interviews' between bell hooks and Gloria Watkins towards the end of her latest book, *Yearning* (hooks, 1990).
3 The 'persuasive fictions' of postmodern ethnography, such as those collected by Clifford and Marcus (1986), have been roundly criticized for their inadequate representation of feminist voices. See, for example, the series of 'cautions' about the postmodernist turn in anthropology issued by Mascia-Lees, Sharpe and Cohen (1989).
4 In his discussion of postmodernism as 'the cultural logic of late capitalism', Fredric Jameson makes a similar call for new maps that will enable us to grasp our positioning as subjects and help us regain a capacity to act and struggle (1984:92).

REFERENCES

Anderson, K. 1991: *Vancouver's Chinatown: racial discourse in Canada, 1875–1980*. Montreal and Kingston: McGill – Queen's University Press.
Anderson, K. and Gale, F. (eds) in press: *Inventing Places: studies in cultural geography*. South Melbourne: Longman Cheshire.
Asad, T. (ed.) 1973: *Anthropology and the Colonial Encounter*. London: Ithaca Press.
Barnes, T. J. and Duncan, J. S. (eds) 1992: *Writing Worlds: discourse, text and metaphor in the representation of landscape*. London and New York: Routledge.
Berger, J. 1972: *Ways of Seeing*. Harmondsworth: Penguin.
Berman, M. 1982: *All that is Solid Melts into Air: the experience of modernity*. New York: Simon & Schuster.
Bhabha, H.K. (ed.) 1990: *Nation and Narration*. London: Routledge.
Bourdieu, P. 1988: *Homo Academicus*. Cambridge: Polity Press.
Bourne, J. and Sivanandan, A. 1980: Cheerleaders and ombudsmen: the sociology of race relations in Britain. *Race and Class*, 21, 331–52.
Burgess, J. A., Limb, M. and Harrison, C. M. 1988a: Exploring environmental values through the medium of small groups. Part 1: theory and practice. *Environment and Planning A*, 20, 309–26.
Burgess, J. A., Limb, M. and Harrison, C. M. 1988b: Exploring environmental values through the medium of small groups. Part 2: illustrations of a group at work. *Environment and Planning A*, 20, 457–76.
Clifford, J. 1983: On ethnographic authority. *Representations* 1, 118–46. Reprinted in J. Clifford (ed.) 1986: *The Predicament of Culture*, Cambridge, MA: Harvard University Press.
Clifford, J. and Marcus, G. E. (eds) 1986: *Writing Culture: the poetics and politics of ethnography*. Berkeley, CA: University of California Press.
Cocks. J. 1989: *The Oppositional Imagination: feminism, critique and political theory*. London: Routledge.
Connor, S. 1989: *Postmodernist Culture: an introduction to theories of the contemporary*. Oxford: Basil Blackwell.

Cosgrove, D. E. 1985a: Prospect, perspective and the evolution of the landscape idea. *Transactions, Institute of British Geographers*, NS 10, 45–62.

Cosgrove, D. E. 1985b: *Social Formation and Symbolic Landscape*. London: Croom Helm.

Cosgrove, D. E. 1990: Environmental thought and action: pre-modern and post-modern. *Transactions, Institute of British Geographers*, NS 15, 344–58.

Cosgrove, D. E. and Daniels, S. J. (eds) 1988: *The Iconography of Landscape*. Cambridge: Cambridge University Press.

Crush, J. 1991: Progressive human geography: South African texts/contexts. *Progress in Human Geography*, 15, 395–414.

Deutsche, R. 1990: Men in space. *Strategies: A Journal of Theory, Culture and Politics*, 3, 130–7.

Driver, F. 1992: Geography's empires: histories of geographical knowledge. *Environment and Planning D: Society and Space*, 10, 23–40.

FitzSimmons, M. 1989a: Reconstructing nature. *Environment and Planning D: Society and Space*, 7, 1–3.

FitzSimmons, M. 1989b: The matter of nature. *Antipode*, 21, 106–20.

Gates, H. L. Jr. (ed.) 1986: *'Race', Writing, and Difference*, Chicago and London: University of Chicago Press.

Geertz, C. 1983: *Local Knowledge*. New York: Basic Books.

Gregory, D. 1980: The ideology of control: systems theory and geography. *Tijdschrift voor Economische en Sociale Geografie*, 71, 327–42.

Gregory, D. 1989: Areal differentiation and post-modern human geography. In D. Gregory and R. Walford (eds), *Horizons in Human Geography*, London: Macmillan, 67–96.

Hall, S. 1989: The meaning of New Times. In S. Hall and M. Jacques (eds), *New Times: the changing face of politics in the 1990s*, London: Lawrence & Wishart, 116–34.

Harding, S. 1986: *The Science Question in Feminism*. Ithaca, NY: Cornell University Press.

Harding, S. (ed.) 1987: *Feminism and Methodology*. Milton Keynes: Open University Press.

Hartsock, N. C. M. 1983: The feminist standpoint. Reprinted in S. Harding (ed.) 1987: *Feminism and Methodology*, Milton Keynes: Open University Press, 157–80.

Harvey, D. 1973: *Social Justice and the City*. London: Edward Arnold.

Harvey, D. 1989: *The Condition of Postmodernity*. Oxford: Basil Blackwell.

Harvey, D. 1990: Between space and time: reflections on the geographical imagination. *Annals of the Association of American Geographers*, 80, 418–34.

hooks, b. 1990: *Yearning: race, gender, and cultural politics*. Toronto: Between the Lines.

Jackson, P. (ed.) 1987: *Race and Racism*. London: Allen & Unwin.

Jackson, P. 1989a: *Maps of Meaning*. London: Unwin Hyman.

Jackson, P. 1989b: Challenging racism through geography teaching. *Journal of Geography in Higher Education*, 13, 5–14.

Jackson, P. 1991: The cultural politics of masculinity: towards a social geography. *Transactions, Institute of British Geographers*, NS 16, 199–213.

Jackson, P. 1992: The racialization of labour in post-war Bradford. *Journal of Historical Geography*, 18, 190–209.

Jameson, F. 1984: Postmodernism, or the cultural logic of late capitalism. *New Left Review*, 146, 53–92.

Johnston, R. J. and Taylor, P. J. (eds) 1986: *A World in Crisis?: geographical perspectives*. Oxford: Basil Blackwell.

Kolodny, A. 1984: *The Land Before Her: fantasy and experience of the American frontiers, 1630–1860*. Chapel Hill, NC: University of North Carolina Press.

Marcus, G. E. and Fischer, M. M. J. (eds) 1986: *Anthropology as Cultural Critique*. Chicago: Chicago University Press.

Mascia-Lees, F. E., Sharpe, P. and Cohen, C. B. 1989: The postmodernist turn in anthropology. *Signs: Journal of Women in Culture and Society*, 15, 7–33.

Merchant, C. 1982: *The Death of Nature: women, ecology and the scientific revolution*. San Francisco: Random House.

Nicholson, L. J. (ed.) 1990: *Feminism/Postmodernism*. New York and London: Routledge.

Norwood, V. and Monk, J. 1987: *The Desert is No Lady: south-western landscapes in women's writing and art*. New Haven, CT: Yale University Press.

Palmer, B. 1990: *Descent into Discourse: the reification of language and the writing of social history*. Philadelphia: Temple University Press.

Pollock, G. 1988: *Vision and Difference: femininity, feminism and the histories of art*. London and New York: Routledge.

Prince, H. C. 1988: Art and agrarian change, 1710–1815. In D. Cosgrove and S. Daniels (eds), *The Iconography of Landscape*, Cambridge: Cambridge University Press, 98–118.

Redclift, M. 1987: *Sustainable Development: exploring the contradictions*. London: Methuen.

Relph, E. C. 1981: *Rational Landscapes and Humanistic Geography*. London: Croom Helm.

Richards, P. 1985: *Indigenous Agricultural Revolution: ecology and food production in West Africa*. London: Hutchinson.

Said, E. W. 1978: *Orientalism*. New York: Pantheon Books.

Sayer, A. 1989: The 'new' regional geography and problems of narrative. *Environment and Planning D: Society and Space*, 7, 253–76.

Smith, N. 1984: *Uneven Development: nature, capital and the production of space*. Oxford: Basil Blackwell.

Soja, E. W. 1989: *Postmodern Geographies: the reassertion of space in critical social theory*. London: Verso.

Spivak, G. C. 1987: *In Other Worlds: essays in cultural politics*. New York and London: Methuen.

Spivak, G. C. 1988: Can the subaltern speak?. In C. Nelson and L. Grossberg (eds), *Marxism and the Interpretation of Culture*, Urbana, IL: University of Illinois Press, 271–313.

Stallybrass, P. and White, A. 1986: *The Politics and Poetics of Transgression*. London: Methuen.

Williams, R. 1980: The idea of nature. In *Problems in Materialism and Culture*, London: Verso, 67–85.

Wilson, W. J. 1987: *The Truly Disadvantaged: the inner city, the underclass, and public policy*. Chicago: University of Chicago Press.

Women and Geography Study Group/Institute of British Geographers 1984: *Geography and Gender*. London: Hutchinson.

11

Desiderata for Geography: An Institutional View from the United States

Ronald F. Abler

A discipline that does not change with its environment will either become a religion or it will die. I have sometimes been accused of preaching, but I do not consider geography a religion, and since 1984 I have been occupied full-time with disciplinary change and survival, as director of the Geography and Regional Science Program at the US National Science Foundation, as president of the Association of American Geographers (AAG) in 1985–6, and as executive director of the AAG since 1989. The perspective I bring to the subject of this volume – in addition to the American view – is that of a professor-turned-bureaucrat who has tried to monitor and create disciplinary change within American geography's institutional environment. Much of the change that occurs in US science is planned – in broad outline if not in detail. A discipline that can read the pattern of an evolving scientific seascape can, if it is supple and well organized, position itself to advance on the crests of successive waves of intellectual and scientific change. Disciplines that fail to attend closely to the seas around them may well founder in the decades to come.

Because the pace of secular change seems to have increased, disciplines must devote greater effort to monitoring their environments and to marshalling their forces to capitalize on opportunities and avoid hazards. That prospect does not please. I dislike direction, personally and professionally. I am well aware of how difficult it is to assess change and respond to it, especially in a discipline as anarchic as geography. But I have been dragged and driven to conclude that geography's delightful intellectual chaos has become increasingly hazardous to its long-term health. I offer in this chapter my prescription for avoiding irrelevance or obsolescence. I do not claim my recommendations are the only road to health or survival. But I do believe these proposals encapsulate questions the discipline's leaders would do well to address. Failure to do so will leave geography vulnerable to ossification or interment.

A CHANGING WORLD

Three broad trends will dominate the environments of all disciplines for the foreseeable future. First, the declining prosperity of western societies, including the United States, will inevitably affect universities and the specialisms they nurture. Second, sciences will continue to globalize, which means that the environments in which geography operates will increasingly be international and multinational. Third, contrary to popular expectations, the world will continue to become a more complicated place; therefore what geographers have to say about and to the world will become more useful and more valuable.

Economies, Universities and Geography

The United States is no longer wealthy. By world standards it remains well off, even plutocratic. Measured against past levels of disposable income, however, the nation has been impoverished. It faces a steady state at best, and more probably eroding abilities to meet needs its citizens once considered basic. What was personified as Reaganism (Thatcherism in the United Kingdom) has outlived Reagan's term of office and will persist beyond his lifetime. Reduced commitments to public services caused by the mounting costs of maintaining the infrastructure and by declining American competitiveness in the global market-place seem irremediable. Any productivity gains realized seem destined to be neutralized by massive public, corporate and personal debt. One hopes that the financial strains evident in the recession of the early 1990s are temporary, but one searches in vain for evidence to allay the fear that the end of this century marks a major turning point in the country's quality of life.

Unlike earlier economic setbacks, that of the early 1990s seriously affected American universities. The country's colleges and universities were once relatively immune to general economic downturns. As they have grown in size and complexity, and as they have become more dependent on state and federal funding, they have become increasingly vulnerable to economic dislocations, especially those that affect state governments. Sales taxes – the principal revenues for most states – declined sharply during the last quarter of 1989, producing serious revenue shortfalls in two-thirds of the 50 states. State governments facing deficits reduced their commitments to higher education by as much as 30 per cent after 1990 (Cage, 1991). Financial privations are evident among private as well as public universities. An $8.8 million deficit in 1991 and bleak financial projections forced Yale University to consider reducing 'the scope of its scholarly endeavors by paring, and possibly eliminating, several academic departments, including linguistics, sociology, and engineering' (DePalma, 1991:B16). Reviewing programmes to decide which might be trimmed or eliminated is anathema to American geographers, who have seen geography programmes trimmed too often for their liking. But it is the responsible reaction to straitened finances. Across-the-board reductions foster mediocrity by penalizing strong and weak programmes equally.

A discipline that fails to thrive in major academic institutions has historically had difficulty prospering elsewhere, as American geographers know only too well from the long series of difficulties caused by the demise or decline of geography programmes at Harvard, Yale and Stanford in the 1940s and 1950s, and at Michigan, Chicago and Columbia in the 1980s. Presence within leading universities has conferred legitimacy and recognition. Scholarly societies such as the AAG, the Institute of British Geographers and similar national associations provide supra-university coherence and structure to disciplines by publishing journals and organizing annual meetings, but they could not maintain a discipline without their membership bases of university faculty and students. American geography's future, like that of other disciplines, is closely tied to the fortunes of the country's colleges and universities and to the ways those institutions confront the financial challenges of the 1990s (Mather, 1991).

The Globalization of Science

American geography and its hosting universities also face challenges arising from the internationalization of science. In retrospect, remarkable changes have occurred in the scale and organization of geography over the last 50 years. Senior scholars began their careers when regional intellectual networks were more important than national structures. Traversing the country from west to east by train took 60 hours before air transportation was available or affordable. That time distance kept the west coast aloof from the AAG, which always held its meetings in late December and almost always met in midwestern or eastern cities. Cheap air travel and telephone service helped integrate the discipline after 1960, when the AAG began to function as a truly national society. The 1990s will witness an acceleration of the process of nationalization, internationalization and globalization of science and geography that began after World War II.

Many natural sciences have functioned as international entities for decades, with easy and frequent exchanges of ideas and personnel. Global integration remains incomplete, however, even in physics and medicine, although it will not remain so for long. Scarcities of human talent and costs of advanced research facilities are compelling nations to collaborate on the construction and operation of major scientific instruments such as particle accelerators. At the same time, widespread use of computers in scientific research and the ease with which computers can now be networked has given rise to a new view of the scale and scope of science; since 1988, scientific leaders in the USA have been talking about *telescience*, based on broadband computer networks. Ideally, telescience would enable any scientist anywhere to use any scientific instrument (or tap any scientific data base) situated anywhere in the world without leaving his or her home location. Telescience is more than speculation. The US National Science Foundation and the US Department of Defense are investing substantial resources in the computer networks that will make telescience possible.

Universities will be nodes in global telescience networks, and one likely consequence will be increased specialization among the nodes. Economic constraints will make the traditional abundance of specialisms less tenable, and telescience will make it less necessary. Yale may decide to do without sociology and engineering, but it will use the resources it realizes by foregoing those programmes to bolster other specialisms in which it can claim distinction. Whereas at one time a strong American literature programme at Yale would have led other universities to beef up their American literature programmes, that trend may reverse. Other leading universities may question whether they should devote resources to specialisms in which Yale leads. In the future, economic stagnation and electronic access will foster specialization and collaboration. Universities will become more selective in their programmatic investments, and telescience will make the absence of some programmes on campus more palatable than it would have been in the past.

The Global Mosaic

The economic, political and ethnic upheavals of the 1980s and 1990s are forcing Americans to revise their ideas about the rest of the world. Despite the potential for global homogenization inherent in frequent contacts with distant places, localism and regionalism flourish. Many notions underlie the peculiarly American conceit that geography is unimportant, both as substance and as a discipline. One unarticulated but widely held assumption has been that the rest of the world was well on its way to becoming just like the United States. In time, under the influence of beneficent technologies of transportation and communication, the world would become a simpler, homogeneous place, with liberty and McDonalds for all. Who needs geography or geographers in a placeless, spaceless world? But a funny thing happened on the way to the global cultural blender: a lot of places refused to go. On the contrary, many groups have become more fractious, insisting upon linguistic and political autonomy, even in the face of severe economic penalties attached to independence and self-control. The fragmentation of what was once a Soviet empire into antagonistic constituents is the most striking instance of a mosaic dividing into finer pieces, but the process and its underlying causes is a general one. The world will be a more complicated place in the year 2025 than it was in 1975.

Cultural and political fission proceeds in the face of greater economic integration and more intensive uses of the globe's places and spaces. The world's area is finite, but population numbers are not. World population continues to grow, albeit at lower rates than in the past. More important, per capita transaction rates among the peoples of the world are growing faster than population. Greater numbers of people – with their continuous sendings and receivings and restless goings and comings – put more pressure on space, resources, transportation networks and natural systems, fostering congestion and pollution. People throughout the world face ever more pressing needs to understand how places and regions work and how they affect each other – if they hope to keep regional and global systems of ecology, economy and polity working reasonably well. The more intensely an area is used, the greater the needs for analysis, coordination and planning. The world that is abuilding will demand more geographers not fewer, and it will demand more from them.

Geography's current research agenda reads like a catalogue of things people throughout the world need to know.

DESIDERATA FOR AMERICAN GEOGRAPHY IN A CHANGING WORLD

Where does American geography stand with respect to these changes and needs? Geography is not physics or radio astronomy, but it shares some of their attributes and it will be affected by the same accelerating trend toward globalization. Geographers hunger for data about places, and governments and private industry are collecting them, capturing spatially referenced data for use in geographic information systems in huge quantities. In years to come, geographers will access, manipulate and analyse such data at a distance, much as some astronomers now operate radio telescope arrays from afar via computers and broadband communications links. At the moment, physical geography is more fully internationalized than human geography. National and language-realm research traditions remain strong in human geography, as they do in other social sciences. But even human geography has achieved considerable unity in some parts of the world, especially in the North Atlantic English-speaking realm, which Johnston (1991a) treats as a single intellectual community from 1945 onward.

Many geography programmes in the United States lack focus. They emulate at the department level the universities of which they are components by trying to offer courses that cover the discipline and, often enough, the entire world as well. In doing so, they dissipate the discipline's limited stocks of energy and talent. While it is understandable that most geography programmes want to offer courses on Africa and on the geography of cities, doing so means that faculty either teach them as adjuncts to their primary interests or, if their primary interests happen to be Africa or cities, pursue those specialisms without the benefit of like-minded colleagues. Few department would be willing to invest in two or three Africanists or two or three urbanists. As a result, one searches the academic landscape in vain for programmes that are widely known for their depth and excellence. Why is there no eminent clutch of Africanist geographers to which any serious student of Africa, be he or she a geographer or not, would customarily repair for a year's intensive instruction and study? Why is there no similar institute for the geography of cities, or for any of the other topics geographers address? Why does the modal American geographer teach his or her courses in isolation, without the benefits of specialization of task and economies of scale?

The scale and scope of geographic research are also at odds with those of many competitors. The modal form of research in which geographers engage might be termed artisan investigation. Individual scholars study small-scale topics using modest research awards or their own resources. While geographers have pursued science on a craft basis, their colleagues in other specialisms have fomented what might be called an industrial revolution in research. In many natural science specialisms and in medicine, teams of investigators mount coordinated, long-term efforts to address suites of related problems in ways that do achieve both specialization of task and economies of scale. Continuity of effort and scale generate support for doctoral and post-doctoral students, who address manageable pieces of large problems in return for their stipends. Funding organizations favour such projects precisely because of their scope and the synergism they generate by organizing large groups of investigators to address large tasks. If the coordinated, long-term research project represents the industrialization of science, globalization and telescience represent computerization and robotics. As countries might find it possible to skip stages of development, geography may move directly from the artisan scale to that of computerization and global specialization. Doing so, however, or even competing with industrialized science, will require that geographers rethink and revise their ideas about the nature, the scale and the scope of the enterprise in which they are engaged.

Priority for Places and Regions

My generation (I began graduate work in 1963) sold its intellectual birthright for a mess of theoretical and methodological pottage. It strove mightily to distance itself from area studies and regional geography, and in large measure it succeeded. Most geographers in the generation preceding mine sought expertise in one of the world's regions and in a systematic specialism. My generation commonly cultivated one or two topical specialisms, or one topical and one methodological specialism. When we in our turn supervised students, we were uninterested in fostering regional specialization, and most of us would have been incapable of doing so had we tried. Those chickens have come now home to roost.

Geography lacks accomplished scholars to respresent the discipline in critical – and fundamentally geographical – efforts within the institutional structure of American science. The AAG recently determined to increase the representation of geographers on American Council for

Learned Societies (ACLS)–Social Science Research Council (SSRC) Committees. The ACLS and SSRC are both non-profit, non-governmental organizations that play key roles in allocating government and private moneys among disciplines and scholars. In 1990, no geographer served on any of their committees responsible for regional studies, which, according to ACLS reports, were as follows:

Advisory Committee on International Programs
Committee on African Studies
Committee on Chinese Studies
China Grants Selection Committee
Committee on Eastern Europe
East European Language Training Committee
International Research and Exchanges Board (IREX)
Joint Committee on Japanese Studies
Joint Committee on Korean Studies
Joint Committee on Latin American Studies
Joint Committee on the Comparative Study of Muslim Societies
Joint Committee on the Near and Middle East
Joint Committee on South Asia
Joint Committee on Southeast Asia
Joint Committee on Soviet Studies
Joint Committee on Western Europe

The AAG Council developed an embarrassingly short list of geographers to nominate to these committees. In two instances it decided to nominate no one. If there were any American geographers qualified to sit on two critical committees with the best regional specialists other disciplines could muster, the AAG Council overlooked them.

Can this prodigal discipline now go home, reclaim its patrimony, and partake of a regional calf fattening on a world that is growing more complicated every month? I believe it will. As Gregory notes, 'ever since regional geography was declared to be dead – most fervently by those who had never been much good at it anyway – geographers, to their credit, have been trying to revivify it in one form or another' (1978:171). Regionalism persisted as a muted theme in geography after 1950, and recent years have seen a renaissance of regional research (Johnston, 1991b; Johnston, Hauer and Hoekveld, 1991). The regional geography of the 1990s differs from traditional forms, however; it is informed by postmodern thinking that attends carefully to causal structures rooted in gender relationships, political power and social structure. While some geographers may be baffled or dismayed by these perspectives, they cannot gainsay the fact that exciting, penetrating regional analyses are being conducted, studies that will do much to re-establish the discipline's presence among colleagues committed to understanding the places and regions of the globe.

Priority for the Environment

My generation foreswore physical geography and ecological concerns almost as fervently as it abandoned regional geography. Human geography seemed to be where the action was in the 1950s and 1960s. Lingering embarrassment over environmental determinism made it risky for a student in most graduate departments to mention the environment and human activity in the same sentence – or even in the same term paper. Stalwarts like Gilbert White and his students soldiered on, but they were few in number. Luckily, physical geography has resurrected itself in the last 20 years in the United States. In my current position, I encounter far less difficulty in identifying good candidates for earth science and ecological posts than I do when regional expertise is called for.

Robert Kates (1987) suggested that the ecological road still beckons despite geographers' earlier failure to take it. Some time during the coming century, world population will probably stabilize at eight to twelve billion people. What geographers know about earth system states and about the flows of energy and materials among and within the earth's great systems can help provide, on a sustainable basis, the four-fold increase in agricultural output and the six-fold increase in energy production that will be needed to support eight to twelve billion people.

Human as well as physical geographers can make valuable – and increasingly welcome – contributions to a sound ecological future. The global ecological research agendas for the next decade are now being set by the United Nations and international scientific institutions. When the International Council of Scientific Unions (ICSU) met in Sofia in October 1990, it elevated its Special Committee for the International Geosphere Biosphere Project (IGBP) to a permanent Scientific Commission. The general assembly gave 'strong support to regional approaches within IGBP including Regional Research Centers. ICSU intends to pursue cooperative global change activities with the social science community' (ICSU, 1990:1). ICSU also asked its Special Committee on Natural Disasters to 'identify new interdisciplinary research areas, including joint activities with engineers and social scientists' (ICSU, 1990:1). The International Geographical Union (IGU) holds membership in both ICSU and the International Social Science Council (ISSC). It will therefore play a crucial role in defining and achieving the objectives of these worldwide programmes, which focus closely on traditional geographic topics. The 24–9 November 1991 meeting in Vienna of ICSU's International Conference on an Agenda of Science for Environment and Development into the Twenty-first Century (ASCEND 21) was devoted to themes such as growth of resource use, land use and degradation, industry and wastes, energy, health (including environmental impacts), land resources, and quality of life (ICSU, 1991:3–4). Furthermore, IGBP planners have concluded that the most effective structure for the IGBP would be 'a global *system* of [14] *regional networks* dedicated to analysis, research, and training. Each regional network would consist of an array of *regional research sites* and a *regional research center*' (Eddy et al., 1991:vii; emphasis in the original).

Never have earth and natural scientists been more receptive to collaboration with physical geographers and social scientists. Sigma Xi, the Scientific Research Society, is a multidisciplinary honorary society that has long been dominated by natural scientists. Its 15–18 November 1991 Washington, DC, Forum on Global Change and the Human Prospect was subtitled 'issues in population, science, technology, and equity'. The forum was unusual both for its social science content and for the enthusiasm with which participants with backgrounds in the natural sciences welcomed the insights of geographers and social scientists. Never in my experience have geographers been so well placed to help answer the great ecological questions of the day. As Kates (1987) notes, they embrace both physical and social sciences, they bring useful tools such as geographical information systems (GIS) to the tasks at hand, they know how to conduct field investigations, and they are theoretically eclectic.

To prosper in the environment in which they will compete for funds and positions, American geographers would be wise to engineer the synergies that are now latent in the discipline's human and physical components. Physical geography faces a revived challenge within geography, however. Geographers committed to human capital, structurationist and postmodern perspectives are indifferent to physical geography at best and hostile at worst. Their observation that concepts such as nature and natural resources are human constructs freighted with ideological baggage is incontestable. It is true that the abilities of different societies to cooperate with or manipulate environments are constrained by political and power relationships. But the logical conclusion of that line of reasoning is a cultural determinism that would be as extreme as geography's hoary bogeyman of environmental determinism. Physical environments and the scientists who seek to understand them cannot be defined away. Physical environments exist, they must be reckoned with, and a positivist approach yields helpful understandings of their workings.

The physical geography–human geography dichotomy is a facet of the two cultures postulated by Snow (1961), and a way to bridge the different epistemologies and explanatory modes that separate human and physical geographers may be the humanistic premises both share. Most physical geographers are humanists at heart. Climatologists, for example, are deeply concerned 'for the future of society and environment; the implications of increased carbon dioxide and depleted ozone are no small matter. Similarly, biogeographers working in the Amazonian rain forest or North Slope tundra hold a stake in the future of fragile biomes that supersedes their individual research interests' (Marcus, 1992:336). Natural scientists and physical geographers want to improve the human condition, and a basis for a coherent, synergistic disciplinary stance with respect to human and physical geography exists in that shared motivation. Continued separation of human and physical geography would seriously weaken the discipline at precisely the time American society needs a synthetic view of how humanity uses and abuses its physical environment. Human geographers – regardless of their philosophical persuasion – cannot hope to meet that need without the insights of physical geographers, and vice versa.

Priority for the Practical

Refocusing geography to meet the challenges of the 1990s will involve more than rebalancing the topics to which geographers address their research. The new worlds which geographers shall brave demand a complementary redirection of the discipline's focus. Too many geographers still preoccupy themselves with what geography is; too few concern themselves with what they can do for the societies that pay their keep (Sanders, 1986).

Strabo of Amasia, writing in 7 BC, argued in the remarks prefatory to his *Geography* that 'The utility of geography is manifold, not only as regards the activities of statesmen and commanders, but also as regards knowledge both the heavens and of things on land and sea, animals, plants, fruits, and everything else to be seen in various regions' (I.I.1). Strabo's *Geography* reveals a tension between practice and scholarship that persists, 2000 years later. Strabo wrote primarily for the Roman imperial audience. Accordingly, he highlighted information that would appeal to administrators and merchants who conducted affairs throughout the empire. But Strabo also relished knowledge for its own sake. 'A work on geography', he noted, 'involves theory of no mean value' (I.I.19). He larded his descriptions of the Roman world with penetrating explanation and pregnant inference, much of which anticipated formal theories developed in the modern era.

During the last several decades, geographers have reversed the priority Strabo gave to the needs of commanders and statesmen. Too many geographers write only for other geographers. Too few geographers write for people outside the discipline or outside the academy. Without question, an inner-focused conversation about what geographers do with relevant data, methods and theory must be prosecuted vigorously. A specialism that fails to engage in ongoing debate about its fundamental assumptions and procedures would not be a discipline. On the other hand, preoccupation with the inner-focused conversation impedes relations with the world outside the discipline, to its detriment. Geography enjoys no ready-made constituency in society. Its prospective constituency consists of people who are innately curious about places, peoples and interactions among them, and of bureaucrats and business people. That audience is far less interested in what geography is than in the useful knowledge geographers can marshal to satisfy its curiosity, solve practical problems and inform public policy.

Over twenty years ago, I argued that geography was seriously short of well trained practitioners (Abler, Adams and Gould 1971:4). Geographers have partially redressed that imbalance, but they have yet to produce the number of dedicated practitioners they need. Geographers have devoted too much energy and talent to introspection and self-indulgent specialization, and too little to the needs of society. Government officials, leaders in business, and the titans of industry need the advice and insights practising geographers provide (Chorley, 1990; Manning, 1990), and geography needs more people who can advance the discipline among non-academic decision-makers. Addressing fundamental human and ecological needs forthrightly and effectively for five years would do more to improve geography's domestic image than the 35 years American geographers have spent on a quest for respectability via the development of abstract theory.

Priority for Survival

Redirecting American geography's attention toward large social and ecological problems – and toward educating and training a corps of practitioners to confront them – offers an additional benefit: it would help geography survive, and even prosper, during the hard times to come.

American society seems to be ripe for a wave of academic fundamentalism; university faculty face more stringent scrutiny of what they teach, how much they teach and what they study. Derek Bok, then president of Harvard University, opened his 1990 report to the Harvard Board of Overseers as follows:

> Looking back on the 1980s, I am struck by a remarkable contrast in the views expressed about American universities. Throughout this decade, we have repeatedly heard from foreign sources that our system of higher education is the best in the world ... Yet surprisingly, far from praising our universities, critics in this country have attacked them more savagely during the past ten years than at any time in my memory ... These harsh indictments have not been culled from fly-by-night journals or obscure newsletters. The fact that they have appeared so often in reputable publications and have attracted the attention of so many readers compels us to take them seriously. (Bok, 1990:1–2)

Media attention to the falsification of research results in the biological and medical sciences, inappropriate allocations of overhead costs from federal research grants, and debates over political correctness have given the public less than flattering views of academic institutions.

American scholars have not taken their critics seriously. The *Chronicle of Higher Education* (the weekly newspaper that caters to American higher education) carried only cursory notices when *The Hollow Men* and *ProfScam* were published. Charles S. Sykes's *The Hollow Men: politics and corruption in higher education* (1990) relates 'the triumph of ideology over ideas in American higher education' by focusing on racial discord at Dartmouth College. It follows Sykes's 1988 book, *ProfScam*, which bears the subtitle *professors and the demise of higher education*. Another treatise laying similar charges was *Tenured Radicals: how politics has corrupted our higher education* (Kimball, 1990). Few faculty members have read these books. Those who have done so have dismissed them. They err. Sykes's books were favorably reviewed in *The New York Times*, *The Wall Street Journal* and *The Washington Post*. They are being read by educated Americans and, more important, by the legislators and trustees who control the fate of American universities (Trachtenberg, 1989, 1991).

To add to these problems, a report on scientific papers in all disciplines published between 1981 and 1985 alleged that 55 per cent

Table 11.1 Percentage papers never cited, by specialism

Specialism	%
Arts and humanities	98
Social sciences	75
Engineering	72
Mathematics	56
Medicine	46
Geosciences	44
Biological sciences	41
Chemistry	39
Physics	37
Political sciences	90
International relations	83
Language and linguistics	80
Anthropology	80
Sociology	77
Business	77
Archaeology	76

Source: Hamilton, 1991

were *never* cited in the five years following their publication (Hamilton, 1990), suggesting that over half of all academic research is a dead-end effort. Applying the results of an earlier study to that 1981–5 cohort suggests that only a fifth were cited more than once. For every 100 journal items published, one could infer that 55 were never cited by anyone, 26 were cited once and 19 were cited more than once – including self-citations. Considerable variation exists among specialisms (table 11.1). An inquiry to the Institute for Scientific Information (ISI) about geography elicited the information that the two reports in question were based on all materials cited (not research articles alone). The inquiry also revealed that for geography research articles, the uncitedness rate from 1984 to 1988 inclusive was about 50 per cent for those published in 1984, not the 74 per cent rate when maps, books, letters, reports, etc. are included (Pendlebury, 1991).

Citation analysis is not without its faults – especially so far as scholars in the United Kingdom are concerned (Anderson, 1991) – and the two *Science* reports touched off a flurry of criticism and a clarification by ISI ('Letters', 1991). No proof exists that works that are not cited are not read and that they have little influence. But a 50 per cent uncitedness rate for research articles arouses that suspicion, and other questions have been raised about how many people besides geographers read geographical journals (Whitehand, 1984; B. L. Turner, 1988). If few people are influenced by what many geographers and other scientists write, the question of whether society should continue to pay them for writing becomes more than a know-nothing *ad hominem*. It is a legitimate question of how well scarce resources are being used.

Geographers and their colleagues in all specialisms will confront increasingly sharp questions about whether anyone benefits from the research conducted in the nation's universities beyond those employed to produce it. Geography programmes in American universities should be prepared to withstand intense and continuous evaluation throughout the 1990s. That examination will occur not because geography programmes differ in any fundamental way from others in the country's universities, but because the nation can no longer afford the kinds of commitment it has made to tertiary education in the past, and because it is no longer willing to underwrite an educational system it does not understand and does not respect. To the degree that American geography programmes conduct useful research and train students to address practical problems, they will fend off further erosion of the discipline's status and forestall the external interference in their affairs that lies latent in current discontent with American higher education.

American geography will better withstand scrutiny if it resolves its related problems of size, curriculum and scale. Geography remains a small discipline in the United States, one that is easily overlooked and whose programmes are vulnerable to absorption or elimination simply because of its size. The AAG is the smallest of comparable scholarly societies (table 11.2) A recent volume on *The Origins of American Social Science* (Ross, 1991) contains not a word about geography. If they hope to prosper, geographers need to become more numerous and more visible within the institutional structure of American science than they have been in recent years, especially within organizations such as the American Association for the Advancement of Science, the National Academy of Sciences, Sigma Xi and the Social Science Research Council.

For American geography, the related problems of numbers and visibility will be eased by the massive efforts now being mounted to improve school geography in the United States (US Department of Education, 1991), but those solutions will raise another challenge. Developing a progressive and cumulative undergraduate curriculum will be one of the profession's most pressing tasks in the 1990s. Currently, geography

Table 11.2 Individual membership in scholarly societies, 1991

Society	No. of members
American Anthropological Association	12,200
American Economics Association	21,000
American Historical Association	14,000
American Meteorological Society	10,000
American Planning Association	25,000
American Political Science Association	11,500
American Sociological Assocation	12,900
Association of American Geographers	6,200
Ecological Society of America	6,500
Geological Society of America	17,000

Source: Personal communications

programmes in American universities are designed to transform second- or even third-year students into geography majors by the end of the four-year baccalaureate programme. In the past, few or no students matriculated as geography majors. Most were recruited from service courses as sophomores or juniors. Therefore most geography courses – across the undergraduate curriculum – are introductory courses. Few have meaningful prerequisites that are consistently enforced. American undergraduate geography curricula are rarely progressive and cumulative, with intermediate and advanced courses that build sequentially on introductory courses. Changes in that historic pattern are now afoot. Geography programmes will soon face dozens or even hundreds of undergraduates who will arrive at universities with some training in geography and who will want to major in geography. The discipline will have to offer them more than a long sequence of introductory courses if it hopes to maintain their interest and excitement. If it fails to meet that challenge, geography will lose many interested and talented students to better-structured curricula.

Geographers would be wise, in appropriate realms, to enlarge the scale and scope of their research efforts. Artisan-scale science cannot yield the results and the visibility that are obtained by large teams that focus on big questions. The AAG's Comparative Metropolitan Analysis Project drew over 70 of the nation's geographers together in a single effort (Adams, 1976). It lent the specialisms involved a sense of common purpose and unity that was exhilarating, and it produced results that attracted widespread and favourable attention outside the discipline. More recent examples of large, coordinated projects are the projected five-volume *History of Cartography* Project (Harley and Woodward, 1987), the three-volume Historical Atlas of Canada (Harris, 1987; Kerr and Holdsworth, 1990), the *Atlas of Pennsylvania* (Cuff et al., 1990), the three-year research project on the national and global structure of manufacturing industries (Technopoles of Southern California) led by Allen Scott at UCLA with support to the tune of $223,000 from the National Science Foundation, and the Earth Transformed project (B. L. Turner et al., 1990), the genesis of the George Perkins Marsh Institute (1991) that was recently established at Clark University. Until such long term, multi-investigator, collaborative research becomes common in geography, the discipline will forego the many benefits resulting from comprehensive research programmes.

Priority for Priorities?

Exhorting colleagues to restructure their professional lives comes easy. Getting them to do so is difficult, for it raises explosive questions of priorities and who sets them. I offer no solutions to that quandary, which will be especially refractory in geography. The AAG hosts 41 specialty groups (table 11.3). I would not want to tell any one of them that its interest will occupy a secondary or tertiary position on the discipline's research agenda for the next decade. Nor would I want to decide which university programmes should focus on which specialisms. As unpalatable as that prospect is, it is one the discipline would be prudent to face. Priority setting is in the air in the United States, just as surely as the financial strictures that are one (but only one) of its causes (Koshland, 1991). An initial and less than satisfactory attempt to set priorities in the social sciences (Gerstein et al.) appeared in 1988. A conference of the nation's scientific leaders was called at the National Academy of Sciences in December 1991 to review a draft report on priority-setting across and within the sciences (Government–University–Industry Research Roundtable, 1991). With few exceptions, the 250-plus conference participants – the top leaders of funding agencies, scientific societies, private corporations that engage in research, and universities – accepted the fundamental assumption of the discussion document and the conference speakers:

> that without priority-setting at the national level, there will be increasing confusion about and less than optimal investments in

Table 11.3 AAG specialty group membership, 1991

Group	No. of members
Africa	170
Aging and the Aged	45
American Ethnic Geography	New
American Indians	71
Applied	340
Asian	184
Bible	43
Biogeography	241
Canadian Studies	113
Cartography	467
China	121
Climate	289
Coastal and Marine Geography	132
Contemporary Agriculture and Rural Land Use	145
Cultural Ecology	211
Cultural Geography	216
Energy and Environment	228
Environmental Perception and Behavior	265
Geographic Information Systems	706
Geographic Perspectives on Women	153
Geography in Higher Education	174
Geography of Religion and Belief Systems	New
Geomorphology	339
Hazards	123
Historical Geography	377
Industrial Geography	192
Latin America	247
Mathematical Models and Quantitative Methods	252
Medical	137
Microcomputer	302
Political Geography	306
Population Geography	248
Recreation, Tourism, and Sport	214
Regional Development and Planning	319
Remote Sensing	344
Rural Development	180
Socialist Geography	133
Soviet, Central, and East European	151
Transportation	186
Urban	569
Water Resources	290

Source: AAG records

> frontier research and research infrastructure of vital importance to the nation . . . Government leaders must set broad national priorities for research in consultation with the individual scientific and engineering disciplines, the larger scientific community, academic institutions, and industry. (pp. 13–16)

Priority-setting seems inevitable. In such an environment, geographers can set their own priorities or let others, less qualified to do so, set priorities for them.

A TURNING POINT

American geography now faces another of the periodic metamorphoses that have twice in the past set its direction for a generation. The early 1990s will be seen in retrospect as a period of fundamental change comparable to the rejection of environmental determinism in the 1920s and to the shift to spatial analysis (the so-called quantitative revolution) of the 1950s. Like the earlier reconstructions of the discipline, the one now under way occurs during a period of rapid turnover. The generation of geographers who completed doctorates in the years following World War II have recently retired. Geography's rising popularity on university campuses is generating net gains of faculty positions despite these retirements. The current intellectual (r)evolution has many facets that are explored in detail in other chapters in this volume, but in general it pits positivist human and physical geographers against a younger group of geographers whose research and teaching are heavily influenced by combinations of phenomenology, human capital and radical perspectives, structurationist views and deconstruction. Positivists view the world as ultimately knowable through the application of scientific analysis. Those who suspect or reject positivism argue that the worlds that are worth knowing can be understood only by attending carefully to the political and social relationships (including gender) that geographers of a positivist persuasion usually ignore. Pickles and Watts (1992:30) contend that 'geographers have begun to realize more clearly how their concepts, theoretical frameworks, methodologies, categories, and language arise out of a particular historical and spatial conjuncture – modernity. Behind this recognition is also acceptance of the need to rethink many of the approaches they use to deal with the world.'

The American geographers who rejected environmental determinism in the 1920s also rejected the quest for theory (especially grand theory) and the statistical methods they associated with proponents of determinism such as Ellsworth Huntington (1945). They turned to regional geography, often at a micro-scale. Thirty-five years later another generation of geographers became enchanted with the theory and quantification its elders had foresworn. Esau sold his birthright because he was hungry. Geographers of my generation hungered for method, theory and philosophy. In acquiring them, they enriched the discipline, but at the expense of cultural geography and area studies, to the degree that the discipline now lacks regional specialists at a time when they are in great and growing demand (Spencer, 1984; Swearingen, 1984; Knapp, 1985; Sagers and Demko, 1985). Will American geographers now reject positivist analysis and physical geography only to discover in the year 2025 that they need and want them?

The history of previous transformations suggests that they will. Without deliberate efforts to set another course, regional geography will remain a poor cousin to systematic geography despite the interest of the postmodern group in local and regional processes. Agencies that underwrite geographic research – especially the National Science Foundation – favour projects that advance theory and methods. Deeper understandings of places and regions, and support for the long periods of language study and fieldwork prerequisite to acquiring regional expertise, rank low among their priorities. Absent strenuous efforts to the contrary, physical geographers and human geographers will continue to go separate ways, to the detriment of both and of the discipline's ability to proffer sound advice regarding human use of the earth. In addition, without programmes to bridge the gap, those who profess geography and those who practise it will continue to talk past each other.

Whether the overreactions of the past will be repeated depends upon the wisdom of geographers who completed their doctorates during the last 15 years. A major philosophical and methodological shift *is* under way in geography. There can be little doubt as to its outcome – youth will triumph. Intellectual changes do not result from conversions of individuals from old ideas to new ones, even in non-paradigmatic specialisms such as geography. Disciplines evolve because of births and deaths. Adherents of what was once a dominant view eventually become a minority, and then a nonentity, as they are replaced by younger scholars reared in different intellectual traditions (Kuhn, 1962). Wide swings in disciplinary perspective and philosophy may be inevitable; wholesale abandonment or rejection of the traditional may be prerequisite to innovation. But I hope that is not so in this instance, and I hope the next generation of geographers will profit from the mistakes made by its two predecessors.

An Impossible Science?

A close intellectual neighbour, American sociology, was recently characterized from within as an *impossible* science (Turner and Turner, 1990) because of its failure to become an integrated discipline. A science-oriented establishment that emphasized methods and theory dominated sociology in the 1950s and 1960s, but it was never able to marginalize competing perspectives. 'The field continues to generate regional associations, subspecialties ("sociologies of . . ."), and specialized journals to publish the research of groups of scholars who may have a lot in common with each other but relatively little in common with other groups' (Berger, 1990:1020). There is little likelihood for improvement in its status because:

> Sociology is driven by fluctuating sources of support – private foundations, the federal government, student numbers, social problems, university politics – that are likely to sustain it as the mixed, and vulnerable, intellectual enterprise it is, without strong ties to central institutions or powerful independent professions of the society, such as political science has with law, economics has with business, or anthropology traditionally has had with the colonial policies of imperialist nations. (Berger, 1990:1020)

Sound familiar?

A Possible Science

Although the parallels between geography and sociology are perplexing, geographers can deliver themselves from impossibility. American geography has matured over the last several decades. Geographers are reclaiming their regional birthright, and as American policy-makers come to grips with the global economy and with broad-scale ecological and political processes, they are rediscovering the value of regional expertise. Geographers have achieved a better balance between physical and human geography than has existed for most of the period since World War II. They are claiming another intellectual birthright that should place the discipline in a good position to address the ecological problems that will command public and scholarly attention for the foreseeable future. On the human side of the discipline, geographers are addressing a broader set of concerns than the economic and urban topics that dominated the literature through the early 1980s. Postmodern perspectives have enlarged and enriched the discipline's stock of variables that explain local and regional variety. Geographers are paying more attention to synthesis. They show a greater interest in telling a coherent tale about how a place, a region or a geographical system evolved or works, as opposed to the post-war emphasis on abstract analysis.

The discipline faces daunting challenges, but meeting them does not lie beyond the abilities of American geographers. I believe that a possible and an enduring science lies within their grasp. I can offer no guarantee that the disciplinary changes prescribed here will deliver geography from the fate that has befallen sociology. But geography, unlike sociology, possesses strong traditions that arise from the fundamental and abiding questions geographers have posed and answered for two thousand years. I see no reason why geography cannot regain, as it begins a third millennium as a self-conscious discipline, the coherence that it would achieve by focusing on places and regions, on humankind's stewardship of the earth, and on the problems upon which survival not only as a discipline – but as a species – depends.

ACKNOWLEDGEMENTS

I gratefully acknowledge the improvements to this chapter suggested by Dr Barbara R. Bailey. A partial and preliminary version of this chapter was published as 'The prospect for Geography in the 1990s' in *Seoul Journal of Education*, 1, 1–11.

REFERENCES

Abler, Ronald F., Adams, John S. and Gould, Peter R. 1971: *Spatial Organization: the geographer's view of the world*. New York: Prentice-Hall.

Abler, Ronald F., Marcus, Melvin G. and Olson, Judy M. (eds) 1992: *Geography's Inner Worlds: pervasive themes in contemporary American geography*. AAG Occasional Publication No. 2. New Brunswick, NJ: Rutgers University Press.

Adams, J. S. (ed.) 1976: *Urban Policy Making and Metropolitan Dynamics*. Cambridge, MA: Ballinger.

Anderson, Alun. 1991: No citation analyses please, we're British. *Science*, 3 May, 639.

Berger, Bennett M. 1990: The offenses of sociology: review of *The Impossible Science: an institutional analysis of American sociology*, by Stephen Park Turner and Jonathan H. Turner. *Science*, 16 November, 1020–1.

Bok, Derek 1990: *The President's Report 1988–1989*. Cambridge, MA: Harvard University.

Cage, Mary Crustal 1991: 30 states cut higher-education budgets by an average of 3.9% in fiscal 1990–91. *The Chronicle of Higher Education*, 26 June, A1–A17.

Chorley, Roger 1990: A blue-print for the future. *The Geographical Journal*, 156, 3, 323–9.

Cuff, David J., Young, William J., Muller, Edward K., Zelinsky, Wilbur and Abler, Ronald F. (eds) 1989: *The Atlas of Pennsylvania*. Philadelphia: Temple University Press.

DePalma, Anthony 1991: With deficit, can Yale still be great?. *New York Times*, 4 December, B16.

Eddy, J. A., Malone, T. F., McCarthy, J. J. and Rosswall, T. (eds) 1991: *Global Change System for Analysis, Research, and Training (START)*. Report of a meeting in Bellagio, 3–7 December 1990. Boulder, CO: IGBP.

'George Perkins Marsh institute established at Clark University' 1991: *AAG Newsletter*, 26, 14.

Gerstein, Dean R., Luce, R. Duncan, Smelser, Neil J. and Sperlich, Sonja (eds) 1988: *The Behavioral and Social Sciences: achievements and opportunities*. Washington, DC National Academy Press, Committee on Basic Research in the Behavioral and Social Sciences.

Government–University–Industry Research Roundtable 1991: *Fateful Choices: the future of the U.S. academic research enterprise. A discussion paper*. Draft dated 25 November. Washington, DC: National Academy of Sciences.

Gregory, Derek 1978: *Ideology, Science and Human Geography*. London: Hutchinson.

Hamilton, David P. 1990: Publishing by – and for? – the numbers. *Science*, 7 December, 1331–2.

Hamilton, David P. 1991: Research papers: who's uncited now?. *Science*, 4 January, 25.

Harley Brian, J. and Woodward, David (eds) 1987: *History of Cartography: cartography in prehistoric, ancient, and medieval Europe and the Mediterranean*. Chicago: University of Chicago Press.

Harris, R. Cole (ed.) 1987: *Historical Atlas of Canada I: from the beginning to 1800*. Toronto: University of Toronto Press.

Huntington, Ellsworth 1945: *Mainsprings of Civilization*. New York: Mentor Books.

ICSU (International Council of Scientific Unions) 1990: Summary highlights. 23rd ICSU General Assembly, Sofia, Bulgaria, 1–5 October.

ICSU (International Council of Scientific Unions) 1991: International Conference on an Agenda for Science for Environment and Development into the 21st Century (ASCEND 21). 21 February. Paris: ICSU.

Johnston, R. J. 1991a: *Geography and Geographers: Anglo-American human geography since 1945* (fourth edition). New York: Wiley.

Johnston, R. J. 1991b: A place for everything and everything in its place. *Transactions, Institute of British Geographers*, NS 16, 131–47.

Johnston, R. J., Hauer, J. and Hoekveld, G. A. (eds) 1991: *Regional Geography: current developments and future prospects*. New York: Routledge.

Kates, Robert W. 1987: The human environment: the road not taken, the road still beckoning. *Annals of the Association of American Geographers*, 77, 525–34.

Kerr, Donald and Holdsworth, Deryck W. (eds) 1990: *Historical Atlas of Canada III: addressing the twentieth century*. Toronto: University of Toronto Press.

Kimball, Roger 1990: *Tenured Radicals: how politics has corrupted our higher education*. New York: Harper and Row.

Knapp, Gregory 1985: Geography and research manpower needs for Latin America and the Caribbean. *Latin American Studies Association Forum*, 16, 19–21.

Koshland, Daniel E., Jr. 1991: The best of times, the worst of times. *Science*, 31 May, 1229.

Kuhn, T. S. 1962: *The Structure of Scientific Revolutions*. Chicago: University of Chicago Press.

'Letters' 1991: Science, 22 March, 1408–11.

Manning, Edward 1990: Presidential address: sustainable development, the challenge. *The Canadian Geographer*, 34, 290–302.

Marcus, Melvin G. in press: Humanism and science in geography. In Ronald F. Abler, Melvin G. Marcus and Judy M. Olson (eds), *Geography's Inner Worlds: pervasive themes in contemporary American geography*, AAG Occasional Publication No. 2, New Brunswick, NJ: Rutgers University Press, 327–41.

Mather, John R. 1991: President's column. *AAG Newsletter*, 26, 1–2.

Pendlebury, David A. 1991: Personal communication, 20 February.

Pickles, John and Watts, Michael 1992: Paradigms for inquiry? In Ronald F. Abler, Melvin G. Marcus, and Judy M. Olson (eds), *Geography's Inner Worlds: pervasive themes in contemporary American geography*, AAG Occasional Publication No. 2, New Brunswick, NJ: Rutgers University Press, 301–26.

Ross, Dorothy 1991: *The Origins of American Social Science*. New York: Cambridge University Press.

Sagers, Matthew J. and Demko, George J. 1985: The nonreplacement of senior scholars in the Soviet/East European specialty in geography. *Soviet Geography*, 26, 199–209.

Sanders, Ralph 1986: Personal communication, 19 March.

Snow, C. P. 1961: *The Two Cultures and the Scientific Revolution*. New York: Cambridge University Press.

Spencer, J. E. 1984: Southeast Asia. *Progress in Human Geography*, 8, 284–8.

Strabo 1917: *Geography*. London: Heinemann.

Swearingen, Will D. 1984: Foreign languages and the terrae incognitae. *Professional Geographer*, 36, 73–5.

Sykes, Charles J. 1988: *ProfScam: professors and the demise of higher education*. Washington, DC: Regnery Gateway.

Sykes, Charles J. 1990: *The Hollow Men: politics and corruption in higher education*. Washington, DC: Regnery Gateway.

Trachtenberg, Stephen J. 1989: From whence comes *ProfScam*?. *Association of Governing Boards Reports*, 31, 30–3.

Trachtenberg, Stephen J. 1991: Academia under indictment. *Association of Governing Boards Reports*, 33, 25–9.

Turner, B. L., II 1988: Whether to publish in geography journals. *The Professional Geographer*, 40, 15–18.

Turner, B. L., II, Clark, W. C, Kates, R. W., Richards, J. F., Mathews, J. T. and Meyer, W. B. (eds) 1990: *The Earth as Transformed by Human Action: global and regional changes in the biosphere over the past 300 years*. Cambridge: Cambridge University Press.

Turner, Stephen Park and Turner, Jonathan H. 1990: *The Impossible Science: an institutional analysis of American sociology*. Newbury Park, CA: Sage Publications.

US Department of Education 1991: *America 2000: an education strategy*. Washington, DC: USGPO.

Whitehand, J. W. R. 1984: The impact of geographical journals: a look at ISI data. Area, 16, 185–7.

The Contributors

RONALD F. ABLER is Professor of Geography at Pennsylvania State University and Executive Director of the Association of American Geographers.

PETER DICKEN is Professor of Geography at the University of Manchester.

A. S. GOUDIE is Professor of Geography at the University of Oxford.

PETER JACKSON is Lecturer in Geography at University College London.

R. J. JOHNSTON is Vice-Chancellor of the University of Essex and was formerly Professor of Geography at the University of Sheffield.

MARTIN PARRY is Director of the Environmental Change Unit at the University of Oxford.

I. G. SIMMONS is Professor of Geography at the University of Durham.

GRAHAM SMITH is Lecturer in Geography at the University of Cambridge.

SUSAN J. SMITH is Professor of Geography at the University of Edinburgh.

PETER J. TAYLOR is Professor of Geography at the University of Newcastle upon Tyne.

Index

economic recession: impact on university funding, 152, 216–17, 229; role of state in, 17
economy, global, 9, 31–2; causes of change in, 34–47; declining US role in, 86–7; indicators of change in, 32–4; research agenda on, 47–50;

Index compiled by Ann Barham

Related Titles: List of IBG Special Publications

1 Land Use and Resource: Studies in Applied Geography (A Memorial to Dudley Stamp)
2 A Geomorphological Study of Post-Glacial Uplift
John T. Andrews
3 Slopes: Form and Process
D. Brunsden
4 Polar Geomorphology
R. J. Price and D. E. Sugden
5 Social Patterns in Cities
B. D. Clark and M. B. Gleave
6 Fluvial Processes in Instrument Watersheds: Studies of Small Watersheds in the British Isles
K. J. Gregory and D. E. Walling
7 Progress in Geomorphology: Papers in Honour of David L. Linton
Edited by E. H. Brown and R. S. Waters
8 Inter-regional Migration in Tropical Africa
I. Masser and W. T. S. Gould with the assistance of A. D. Goddard
9 Agrarian Landscape Terms: A Glossary for Historical Geography
I. H. Adams
10 Change in the Countryside: Essays on Rural England 1500–1900
Edited by H. S. A. Fox and R. A. Butlin
11 The Shaping of Southern England
Edited by David K. C. Jones
12 Social Interaction and Ethnic Segregation
Edited by Peter Jackson and Susan J. Smith
13 The Urban Landscape: Historical Development and Management (Papers by M. R. G. Conzen)
Edited by J. W. R. Whitehand
14 The Future for the City Centre
Edited by R. L. Davies and A. G. Champion
15 Redundant Spaces in Cities and Regions? Studies in Industrial Decline and Social Change
Edited by J. Anderson, S. Duncan and R. Hudson